PERCEPTION

RECEPTION

HOW OUR BODIES

PERCEPTION

SHAPE OUR MINDS

DENNIS PROFFITT AND DRAKE BAER

ST. MARTIN'S PRESS ⚏ NEW YORK

First published in the United States by St. Martin's Press,
an imprint of St. Martin's Publishing Group

www.stmartins.com

Designed by Richard Oriolo

Library of Congress Cataloging-in-Publication Data

Names: Proffitt, Dennis, author. | Baer, Drake, author.
Title: Perception : how our bodies shape our minds / Dennis Proffitt and
    Drake Baer.
Description: First edition. | New York : St. Martin's Press, [2020] |
    Includes bibliographical references and index.
Identifiers: LCCN 2020012802 | ISBN 9781250219114 (hardcover) |
    ISBN 9781250219121 (ebook)
Subjects: LCSH: Perception. | Somesthesia. | Mind and body.
Classification: LCC BF311 .P7476 2020 | DDC 153.7—dc23
LC record available at https://lccn.loc.gov/2020012802

Our books may be purchased in bulk for promotional, educational, or business
use. Please contact your local bookseller or the Macmillan Corporate and
Premium Sales Department at 1-800-221-7945, extension 5442, or by
email at MacmillanSpecialMarkets@macmillan.com.

First Edition: 2020

10 9 8 7 6 5 4 3 2 1

For our families

CONTENTS

Of all things the measure is man, of the things that are, that they are, and of the things that are not, that they are not.
—Protagoras of Abdera, 490–420 BCE

To see the organism in nature, the nervous system in the organism, the brain in the nervous system, the cortex in the brain is the answer to the problems which haunt philosophy.
—John Dewey, *Experience and Nature,* 1925

PERCEPTION

INTRODUCTION

I SING THE BODY ELECTRIC

F YOU'VE ENJOYED THE ENDURING PLEASURES of a romantic relationship, then you may have noticed that your partner—we'll call them Sweetie— has a particular, appealing smell. In the air about them or the scent of their shirt, there is a compelling, individual allure. One of the pleasures of a kiss is the accompanying whiff of Sweetie's aroma. It turns out that body odor is as unique as a fingerprint. No two people smell exactly alike. This is how bloodhounds can track Sweetie over terrain that has been traversed by dozens of other people. What makes Sweetie's smell unique is their major histocompatibility complex (MHC), which is the large group of genes that codes for the immune system. There is enormous variability in human MHC, and everyone's individual complex is unique. MHC is present in sweat as are a number of other chemicals, such as those derived from the foods you eat. In romantic relationships, the scent of Sweetie's MHC carries an important message about whether they would make a suitable mate, someone to have children with.

In a study conducted by Swiss zoologist Claus Wedekind, it was found that women perceive as most pleasant and attractive the smell of men whose MHC is maximally different from their own.[1] This is an evolutionary adaptation at work: if you mate with someone whose immune system is similar to your own, then problematic recessive traits have a higher likelihood of being expressed. Increased risk for harmful recessive trait expression is why there's a taboo against mating with near relatives, and why royal inbreeding, from the pharaohs to the Habsburgs, has led to developmental problems. Keeping the blood "pure" actually has unintended risks. Hence why evolution guides you the other way: the benefit of being attracted to someone with a maximally different immune systems is reproductive fitness—you'll likely have healthier children.

But notice how personally subjective this perception is: Sweetie's great smell is a sign of reproductive fitness for you, as an individual, mating with Sweetie. Someone else, with a different MHC, may not find Sweetie's bodily odors to be such an olfactory delight. You are also different from the bloodhound: bloodhounds might pick up on Sweetie's scent, but they're not likely to have the same emotional response that comes from catching a whiff of probable reproductive success. The realm of perceived human body odors is not an objective world of good and bad smells. Rather, you perceive a social world of scent that is as unique as your own immune system. The goodness of Sweetie's smell is due to both your and Sweetie's genetics. It says "Let's get to know each other. We are a good match." Hence smell is a signal of, and force for, bonding, collaboration, and love. But like much of social perception, it is also a medium of, and conduit for, repulsion, otherization, or even hate. When the first Europeans arrived in Japan, their diets—heavy on animal fats—led to body odors that the Japanese found offensively rank and reminiscent of butter—such that the Europeans earned the derogatory label *bata-kusai,* or butter stinkers, a slur for Westerners that remains part of the language to this day.[2]

We develop and are immersed in a human social ecology, and consequently our perceptions are shaped by our upbringing in the cultural environment in which we were raised. We cannot help but perceive people of different skin tones as being of "other races," a bias that emerges well before we learned to speak. At three months of age, infants raised in a homogeneous ethnic environment prefer to look at people of their own group rather than at other ethnicities, but the effect is less pronounced for children who grow up in more diverse social environments. By five months of age, infants prefer to look at someone who speaks their native language; older infants are quicker to accept toys from native-language speakers; and by the time children reach preschool, they prefer native speakers of their own language as friends.

These developmental biases contribute to "the other race effect," a widely studied pattern in which people have a tougher time recalling or recognizing the faces of people who belong to other ethnic groups, as if the social categories put some kind of opaque filter on perception and memory. Our social perception appears to narrow. One experiment with British babies found that Caucasian three-month-olds were equally able to discriminate between African, Caucasian, Middle Eastern, and Chinese faces, but by the time they reached nine months, they only discriminated between different Caucasian faces.[3] The same has been found with Chinese babies.[4] Hence the oft-repeated but no longer socially acceptable saying, "You people all look the same." If your life experiences thus far have led to a narrowing of your social perception, then you're not going to perceive people of different social groups as richly as you would your own. Instead of seeing people of other backgrounds as individuals, we risk seeing them as only members of their social category. Hence,

"You people all look alike." And, as we'll uncover, that "dis-individuating"—not seeing a person as an individual but just as a member of a group—leads to a feeling of "You people all act the same." White or black, liberal or conservative, boomer or millennial.

Mix deindividuation with stereotypes and cultural assumptions about the indelibility of "race" and you get racial bias, the automatic sorting of people according to whatever beliefs you've received from culture about "what they're like." Jennifer Eberhardt, a Stanford psychologist, has shown how racial stereotypes influence perception—for example, when white undergrads were primed by a slideshow of black faces, they more quickly recognized images of weapons and crime-related objects, compared to if they were shown images of white people.[5] In this study, a silhouette of a knife or gun emerged from a blank background, and the college students who saw the black faces perceived the weapons faster. This is uncomfortable stuff. You can be open-minded and well educated, but you're still invariably going to experience a world biased by your personal history.

Nowhere is this bias more evident than in the political scene of the early 21st century, as polarization became the norm across societies, and especially so in the United States. To some, a politician may be viewed as a champion of American values; whereas to others this same political figure may seem the worst sort of ignoramus. How can such divergent opinions develop among reasonable people living in the same world? The answer, we believe, is that due to differences in their personal histories, the individuals within each of these political camps may live in the same physical world of current affairs, but their subjective experiences are worlds apart. And, as we'll explore in detail, it's not just that we live in social worlds of our own making but that all we perceive, down to the steepness of a hill or the size of a glass, depends on who we are as individuals.

While science has been primarily, and rightfully, concerned with objective truths, there has long been an undercurrent of research into subjective experience. The approach that we will take in this book borrows much from Jakob von Uexküll, a Baltic German biologist who was born in 1864 and died in 1944. Though a little-known historical figure today with an unwieldy last name, the scientist hit upon a key concept that captures the scope of the project at hand. Von Uexküll was interested in how different species experience the same physical world. The German language is known for its specificity, and with von Uexküll it really delivered. He distinguished between the *Umgebung,* which is an objective physical environment, and the *Umwelt,* which is a particular animal's experience of that place. For example, Denny and his dog, Lulu, may walk through the same field, but from Lulu's perspective, Denny misses most of the interesting smells. What you experience is your *Umwelt*: a wildflower means something different to you if you're a cow chewing cud, a bee running pollination errands, or a child picking flowers. Science tends to concern itself with the *Umgebung,* and consequently the relativistic perspective of ecological psychology has received far less attention. But there is nothing about the scientific method that precludes studying subjective experience and uncovering consistent, useful truths about perception. The task begins by asking some very basic questions.

Suppose you were trying to understand what it's like to be a particular kind of animal, such as a bird. You would likely infer that judging by its body and the sorts of behaviors that such a body is capable of, a bird's mental life must have a lot do with flying and the many concerns that such a way of life engenders. In general, when trying to understand what it must be like to be a certain animal, we ask: *What kind of animal are they? What kinds of bodies do they have? And with those bodies, what do they do?* These sort considerations are the obvious starting points for understanding an animal's *Umwelt*—the world that they, themselves, live in.

What about the human *Umwelt*? What experiential worlds do we live in as a species, and how do these worlds differ across individuals? The ecological nature of this question has been mostly overlooked in contemporary psychology, in part because we naively assume that we all know what it's like to be human. But unfortunately we are poor judges of our own experience, and it's common sense to believe that we experience the world as it objectively is. This is what social scientists and philosophers call naive realism: taking what we see, smell, hear, and feel at face value.[6] We project our individual mental experience into the world, and thereby mistake our mental experience to be the physical world, oblivious to the shaping of perception by our sensory systems, personal histories, goals, and expectations. For example, you might say *the movie was good*, ascribing an objective quality to the work of art; but it would likely be more accurate to say that *I liked the movie*, describing, instead, your *perception* of it. Even though our naive intuitions are that we see the world as it is, we do not. We see a human *Umwelt*. Moreover, every human is different and consequently has a somewhat different personal *Umwelt*. We are each living our own personal version of *Gulliver's Travels*, where the size and shape of the objects and people we see are scaled to the size of our body and our ability to interact with our surroundings. Our experience of the world informs us about how we fit in the world. Sweetie's great smell says "Love me. We are a good fit."

While we go about our days with the commonsense assumption that everybody experiences the same world, perception research says that experiential reality—the world you see, hear, feel, smell, and taste—is unique to each individual. That a basketball hoop is ten feet high has a very different meaning if you're four foot seven or seven foot four. Here's an example from the research of Jessi Witt, one of Denny's graduate students and now a professor at Colorado State University.[7] Jessi went to Charlottesville's softball league fields, and as games ended, she asked players to indicate, by pointing at one of a set of differently sized circles displayed on a large poster board, which circle was the same size as a softball. She then asked them to report how many hits and at bats they had had in the just completed game. Players were enticed to participate by the offer of a free sports drink. Jessi found that the higher the batting average—hits per at bats—the larger the ball was reported to be. The perceived size of the softball was influenced by the batter's success in hitting it, a finding that corroborates Mickey Mantle's experience after hitting a mammoth home run: "I never really could explain it. I just saw the ball as big as a grapefruit."[8] Similarly, George Scott, who played for the Boston Red Sox, said, "When you're hitting the ball [well], it comes at you looking like a grapefruit. When you're not, it looks like a black-eyed pea."[9] As Denny, his collaborators, and others have shown, golfers putting well see bigger holes,[10] American football placekickers who are kicking well see the uprights wider and crossbar lower,[11] and successful archers see a bigger bull's-eye—as do dart throwers.[12] If you're obese or fatigued, distances look larger than if you're slim or rested.[13] Capable swimmers judge underwater distances to be shorter, as do people wearing swim fins.[14] If you're holding a tool that helps you reach for things—like one that helps you grab a cereal box off the top shelf in a grocery store—then you'll judge distances as shorter.[15] Distances appear shorter if you've just been driving a car than if you've been walking.[16]

Psychologists rarely ask the sorts of questions about people that naturally initiate inquiries into other animals: What sort of animal are we looking at? What kind of body does it have, and consequently what can it do? Instead, we live in the age of the disembodied brain. As Drake knows firsthand from years in digital journalism, a science writer hunting for a headline need only claim that some new remedy or technique "changes your brain." (But, of course, everything that changes experience changes the brain.) CAT scans show up in the criminal courtroom.[17] And *neuro* itself has become a profligate prefix, having long since expanded beyond the life sciences: you can now engage in neuroethics, neuroeconomics, neuromarketing. Predictably, to the neurophilosopher, the self is the brain. To the cognitive scientist, the brain is a computer, making abstract, symbolic calculations. The body, in all this, is irrelevant, save for maybe as a way of transporting your brain from place to place.

Yet, as a current strain of perceptual research is showing, the way we think, feel, and exist is inexorably shaped by our physical being. The body and brain are indivisibly coupled, and this book is a celebration of—and investigation into—that fact. The better we can understand what our bodies are—what they can do, what they need, what they must avoid—the better we can understand ourselves and our lives. To do this, we need to put the brain back in the body.

In 1852, Walt Whitman wrote his most famous poem, "I Sing the Body Electric."[18] Central to the poem is the image of the human body itself as poetry—walking, laughing, grasping—verse made flesh. The electricity of which he spoke is the experience of being alive. Vital, visceral, and vivid. Ennobled, emboldened, embodied. In the following chapters, we will explore the science of the body electric: What kind of body do we have and how does it shape what we *do,* what we *know,* and how we *connect* with others?

DOING

1.

DEVELOPING

FOR REASONS THAT NO ONE REALLY UNDERSTANDS, we have little or no memories about what it was like to be an infant or toddler. And, of course, we can't simply ask infants to report on their experience because they can't yet talk. Tapping into the developing minds of infants and toddlers requires terribly clever researchers asking the right sort of questions. *What kind of bodies do infants and toddlers have? How do their bodies change over time? And with these bodies, what do they do?* The most general finding deriving from this research is that the simple exercise of their developing motor skills teaches children about what the world is and how it works. This is how human knowledge gets off the ground. As children develop, they first discover that the world is filled with gum-able things that, as time goes by, maybe turn out to also be graspable, throw-able, roll-able, or pliable things. No one initially teaches children how to explore their environment. The cradle of knowledge is exercise and play. Children learn to interpret their experience by actively creating it via crawling, walking, falling, and so forth. Moreover, what we learn in childhood forms the foundation for all the new experiences and discoveries that are to come. And, correspondingly, if we're going to run studies that get at what it's like to be a child, we need to design experiments that allow children to roam free.

LEARNING TO WALK

ON ANY GIVEN DAY, New York University's Infant Action Lab is abuzz in pint-size activity. Led by the pioneering developmental psychologist Karen Adolph, the lab has revealed much about how kids come to discover what their bodies can do in the world in which they find themselves. On the day of Drake's visit, the range of work being done spoke to the scope of the team's inquiries. Research assistants were busy coding data that had been sent back from far-flung laboratories around the world: one was studying a traditional cradling practice in central Asia, where babies are swaddled to the point of immobilization in a cradle for 20 some hours a day, up until 18 months of age. (Don't worry; they turn out fine.) Other Action Lab researchers were going into homes and observing the natural activity of tots. Parents were bringing five-year-olds into the lab to see how they handle what Adolph calls the "hidden affordances of everyday objects," as in, how kids learn the opportunities for active play afforded by the objects in their enlarging surroundings, whether it's navigating a Kleenex box or opening a water bottle.

Adolph is tall, slim, and moves about her lab with a certain frenetic precision—the product, one surmises, of decades spent shepherding toddlers, their mothers, and a constant stream of graduate students, postdocs, and undergraduate research assistants. To mix archetypes, she's equal parts fairy godmother and mad scientist, an iconoclast turned matriarch who, by virtue of her area of study, has been forced to invent not only novel hypotheses but the many bizarre, fascinating platforms, tumbling mats, and quasi-jungle gyms needed to test out what small humans are actually doing as they learn to perceive the world and move through it. Her enthusiasm is infectious. What could be more interesting than studying how, through development and play, our species achieves its commonsense understanding of the world?

Enter through the doorway into the lab, round the corner past the computer bay, and you're in a bright, welcoming playroom, what feels like the stage for a high-production-value children's TV show. This is where the action in the Infant Action Lab takes place. The atmosphere is opposite the sterility that the word *laboratory* might conjure up: Adolph's lab is brightly but not oppressively lit. Primary colors abound. Across the broad center of the room lies a recessed strip in the shape and manner of a long-jumper's pit but solid instead of sand. On the near wall stands shelf after shelf of toys. On the far wall hangs what looks like a horizontal ladder, which turns out to be an adjustable set of monkey bars; and standing next to the other walls are apparatuses for which Adolph is perhaps best known in academic circles. Adolph wants to know, not just what infants can do but also what they think they can do. For example, a typical study might investigate the maximum steepness of an incline that a crawling infant can descend without tumbling, the maximum steepness that they will attempt to crawl down, and how their awareness of their own abilities changes with experience. To answer such questions, Adolph employs platforms and ramps, the height and angle of which can be adjusted with great precision. The fruit of a longtime industrial design collaboration, these apparatuses allow for hills and cliffs and gaps to form and disappear, depending on the experiment in question. Every apparatus has shock-absorbing feet, so that if the floor vibrates this way or that—a real danger when the subway rumbles beneath the building—equilibrium can be maintained. On top of that, the entire room is under the surveillance of cameras, so all behavior therein can be captured and later coded. One is reminded of the Transformers movies as Adolph demos her wares—all gears and garage door openers, slopes and bars, feats of sophisticated, yet utterly approachable, engineering. For all the complexity, the lab is a cozy, easy place for babies and parents to be. "Even when baby is in a headset, everybody is happy," she says.

Infant researchers are some of the most creative scientists around—they have to be. Infants don't talk, so you can't ask them questions. Unlike the young adults, who make up most of the participant pools for psychology experiments, babies don't comply with instructions. You have to set up experimental circumstances in which young ones can behave freely and thereby show the experimenter what they know and are capable of doing. When study participants come into Adolph's lab, the process begins with Adolph or a colleague affirming consent with the caregiver. Then it's time to deck the child in whatever technological rigging is necessary: like a tiny head-mounted camera for eye-tracking studies or a weighted vest or Teflon shoes to test how kids compensate for changes to their body's dynamic balance. The actual running of experimental trials is "a coordinated circus," Adolph warns, with the parent standing to the side and a researcher keeping close to the baby in case they fall. If it's off a platform—which happens with scientific regularity in novice walkers—then the researcher has to pluck them out of the air.

In the vernacular of psychology, Adolph strives for "ecological validity": if you set up an experiment to be more like real life, you'll likely get more accurate information about human behavior as it naturally occurs. Consider the matter of a baby's first steps. For more than a century, researchers had assumed that toddlers walked in straight paths, taking the shortest, most efficient route from point A to point B, right? Adolph thought the same, until she stumbled on to a series of results indicating otherwise. For one fateful experiment, the graduate student she was working with wasn't quite handy with the catch-the-infant-out-of-the-air-drill, so they moved the trials from a raised platform to the floor. Then, when asking the tot to walk, the kid did everything but go in a straight line. Adolph sensed she was on to something, so she contacted every living researcher whom she knew had done similar walking studies, and they all reported the same pattern. Everybody threw out more data than they kept, because the babies were so reliably irregular. Wandering about was the norm; a straight line was the exception.

The assumption that infants walk in a straight line speaks to a larger problem in science, Adolph says: experiments are run in a manner that's convenient for the experimenter, like testing locomotion in a straight line. Then theories are constructed for explaining the phenomenon, and these theories become taken as truth. It's a classic example of what William James, one of the fathers of the field, termed the psychologist's fallacy: rather than seeing human behavior as it is, psychologists' views of human nature are biased by their own preconceptions about what it should be.

Karen Adolph is the rare researcher who can set her assumptions aside when the results of an experiment turn out other than expected. When she found that the babies in her studies didn't walk in straight lines, as the research literature suggested they should, she decided to study how infants naturally walked when unconstrained. The results of these studies showed that babies dance their way through their world following their own creative muse as opposed to the dictates of efficiency. One could make infants "walk the plank" and thereby study linear walking, but what would the results mean? This is not how infants naturally walk. If you want to study what infants do, then you need give them the freedom to do what they want. Adolph is the master of this ecological approach to infant development.

One of Adolph's formative moments came when she was first immersing herself in psychology as a college undergraduate. She recalls coming to her adviser distraught, weepy about a lecture on perception, presented in its classical, conventional way. Mainstream accounts of perception, both then and now, resemble the cliché about history being one damn thing after another. For visual perception, that means: first light forms an image on the retina, and then certain features of the image are extracted, sent to the visual cortex for processing, which is followed by still more processing, and so it goes. Listening to or reading such accounts would never suggest that all of this is going on within a living, behaving, goal-directed person.

"This just can't be like this, this can't be how we see," she remembers her undergraduate self saying through tears. Her adviser looked at her and said, "Well, it's not." Then he reached to his bookshelf and said, "Here, read this." He handed her *The Senses Considered as Perceptual Systems,* the second book by James J. Gibson. "To me, it was like discovering religion," Adolph says. "That was my biggest epiphany." It lead her to James and his developmental psychologist wife, Eleanor Gibson, known together affectionately as Jimmy and Jackie.[1] (For the sake of clarity, we'll refer to the pair as Jimmy and Jackie.) In partnership, the Gibsons were revolutionizing how we think about visual perception and its development.

A NEW WAY OF SEEING

COMMON SENSE TELLS US that the world is as it appears, that our experience of the world *is* the world. We see what is there. Every time we stop at a curb rather than stepping out in front of an oncoming bus, we affirm our belief that it is best to believe in the veracity of our experience. This *naive realism* asserts that there is a one-to-one correspondence between the world and our experience of it.

Once examined, however, naive realism fails to jibe with the facts. To begin with, the stimulus information for vision (light's input) is generally assumed to be an image projected on the back of the eye—the retina—where the neural photoreceptors are located. This retinal image is two-dimensional, upside-down, and lacking in the constant properties that are prevalent in perception. For example, the size of an object's projected retinal image varies with distance—the image gets bigger as the object gets closer—and yet our perceptions are of objects having constant sizes. Hold a pencil out at arm's length, then bring it toward your face. Notice that the pencil's appearance gets bigger while its actual size is perceived to remain the same.

So why did you perceive the pencil as having a constant size when its projected size in your eyes varied with distance? A ready answer is that we *know* how big pencils are and that their size does not change when we move them back and forth. In philosophy this answer is called *idealism* and it asserts that the not-quite-accurate retinal image is augmented and enhanced by knowledge and memories acquired through evolution or individual learning experiences. The great 19th century physician and physicist Hermann von Helmholtz coined an expression for this augmentation process, *unconscious inference*. By his account, if you perceive an object to be a pencil, then you will perceive it to have the properties that pencils are known to have.

For the past 150 years, unconscious inference has held sway as the generally agreed upon mechanism by which retinal images are transformed into perceptions of the world. Jimmy Gibson, however, disagreed. He recognized that the problem with the unconscious inference account was that it could never get off the ground; it just wouldn't work. Children would need to already know all of the properties of things in the world before they could perceive things as having these properties.[2] But where does this knowledge come from? How are babies, for example, supposed to learn the correct size of pencils if all they ever see are their projected images, the size of which varies with distance? How are infants supposed to learn what is actually in the world? It's impossible. The whole approach is untenable, Jimmy argued. We need to start over, he said. Like anyone else, however, scientists don't like to begin again, especially if they are late in their careers. Although he accrued a few converts, the Gibsons and their students mostly had to go it alone.

So, how do we start over? Like many scientists, Jimmy's life experiences shaped his worldview. His father was a train conductor, and the young Jimmy had many boyhood reveries standing at the front and rear of his father's trains. Why did the world appear to loom larger as the locomotive hurtled forward, and why did it shrink into a point on the horizon when he stood in the caboose? These experiences taught him that when you move, everything in your visual world also moves in regular ways that specify the surrounding environment. The way that everything appears to move as you move is called optic flow, and it is easily noticed. Recall an evening drive on a country road, for example. A nearby fence will appear to move by quickly, whereas a more distant mountain or hill will seem to move very slowly. These differences in the observed speed of objects that vary with distance actually specify their relative distances—things that move fast are near you; slower-moving things are farther away. Distances can be derived from optic flow—the speed at which things go by us as we drive, run, or walk is a function of how quickly we are going and how far away these things are. Distances need not be inferred because they can be derived directly in the information available to a moving observer. Similarly, the projected size of objects grows or shrinks as they approach or recede from us, respectively, and their actual size can be derived from their changing projected size. For example, the actual size of the pencil can be determined from its changing size as you moved it back and forth. That's the fundamental Gibsonian insight: the information upon which perception is based is not a static, flat, retinal image, but rather the foundation of perception is optic flow, the motion that occurs whenever the observer moves.

Jimmy took a faculty job was at Smith College in 1928, where he stayed until the war effort summoned him in 1942. The young vision researcher was brought into the US Army Air Force. His commanders had questions that the eight-year-old version of himself would have recognized as his own: How do pilots land planes? And how do we help them do it better? For that matter, how does someone even walk from one place to another? Though these questions seemed elementary, the literature from a hundred years of vision research supplied Jimmy with no answers. Up until then, perceptual research had not thought to tackle such questions.

What Jimmy began to discover in the Army Air Force—and which he would expand upon in the book that would make him famous in his field, *The Perception of the Visual World*—is that we don't see some objective world that's measured out to us in centimeters and inches. Rather, what we perceive are what Jimmy called "affordances."

Affordances are the way that you fit into a given situation, or the possibilities for action that surfaces and objects permit for an organism having a particular body and behavioral repertoire. For an able-bodied person, a solid floor affords walking, whereas the surface of a pond does not. A rock affords grasping and throwing so long as it is of a size and weight that it can be grasped and lifted. "The *affordances* of the environment are what it offers the animal, what it *provides* or *furnishes,* either for good or ill," Jimmy wrote. "The verb *to afford* is found in the dictionary, the noun *affordance* is not. I have made it up. I mean by it something that refers to both the environment and the animal in a way that no existing term does. It implies the complementarity of the animal and the environment."[3] Jimmy dubbed his account of perception the "ecological approach." Perception, he said, is an achievement of a living, behaving organism that is actively exploring its environment. These affordances, which are first discovered in infancy, go on to structure our everyday experience, whether it's throwing a ball, making an investment, or deciding whether or not to trust someone.

Jimmy's position asserts that visual information is sufficient—without augmentation from knowledge, memory, or unconscious inferences—so long as the organism is free to move and explore its environment. What you see is a result of what you can do; put more technically, what is seen by a free-roaming organism is the visual products of its own purpose-driven actions.

A walking person will experience how the world moves past her as she walks, while at the same time experiencing the energy expenditure associated with walking. She will discover the visual properties of objects that can be grasped and inclines that can be ascended on foot. She will discover the affordances of her world. This is not naive realism because the affordances that are perceived in the world are specific to an organism's species, its body, behavioral ways of life, and unique individual differences, like life history, goals, and expectations. It is not idealism because the perception of affordances does not rely on preexisting (or a priori) knowledge. Ecological realism asserts that we perceive the world, not as it is but rather as it is for us. This was Jimmy's insight. It remains the stuff of epiphany. This is how the body shapes the mind.

The Gibsonian project was not just Jimmy's but also Jackie's.[4] She was initially his student when he was a young professor; the two met at a graduation garden party at Smith. Jimmy had been tasked with greeting guests and Jackie with serving punch. Structural sexism kept Jackie out of academics for a long while. An anti-nepotism clause at Cornell meant that she couldn't take a faculty job while Jimmy was working there, so she had to work as an unpaid research assistant. She was rejected by several research labs because experiments were thought to be unladylike. Within psychology, the developmental subfield was deemed "women's work" and safe for her to go into.

Jackie's research began at the Cornell's Animal Behavior Farm, where she raised a range of animals for psychological research, including goats, kittens, turtles, and rats. Human infants and nonhuman animals cannot tell you anything about their experience, so their mental lives must be inferred from their behavior in clever behavioral experiments. Jackie partnered with Richard Walk, a Cornell faculty member with crucial laboratory access. Their collaboration would yield one of the most iconic experiments in midcentury psychology.

The 1960 experiment has come to be known by the apparatus that Gibson and Walk built for it: a "visual cliff."[5] While you likely wouldn't want to try the experiment with a baby at home, it's easy enough to build the apparatus. A visual cliff is simply a table that a baby could fall off. Put thick glass over the table and the falling-off part and that's a visual cliff. Just like that, you've created an optical illusion. If one were light enough—like a baby or a kitten—one could crawl out over the cliff and be supported by a transparent surface, like the glass floors designed to delight thrill seekers in some skyscrapers, the Grand Canyon, and the like.

The experiment went like this: Little Johnny (all the test subjects were male) was placed near the edge of the falling-off part of the table. Mom called to him first from the plunging cliff side and then from the shallow side. As Gibson and Walk wrote in *Scientific American* the following year, the optical illusion worked: "Many of the infants crawled away from the mother when she called to them from the cliff side; others cried when she stood there, because they could not come to her without crossing an apparent chasm," they wrote. "The experiment thus demonstrated that most human infants can discriminate depth as soon as they can crawl." Each of the 27 infants happily crawled on the table's shallow side at least once, but just three dared crawl over the apparent abyss.

The nonhuman animals were another story. If they were raised normally, they'd avoid the deep side of the cliff. But some of the kittens had been raised in the dark. These kittens could see fine, but they lacked experience moving about and exploring an illuminated environment. When these kittens were placed on the visual cliff, they would walk over the shallow or deep side with equal frequency. These kittens acted as if they saw nothing wrong with walking off a cliff; however, after just a week of normal experience in lit spaces, these same kittens assiduously avoided the deep side of the visual cliff.

There were—and still are—powerful implications from these research findings. First, there is a difference between being able to see and understanding what is seen. All the kittens in this experiment could see just fine. However, unlike the normally raised kittens, the kittens raised in the dark initially did not understand that walking off a cliff should be avoided. This leads to a second implication: allow a kitten to freely explore its visual world and it will quickly learn the affordance of its environment; for example, it will learn that it can walk on surfaces but not on thin air. The constraints of this experiment, however, left lingering questions for the field. Human babies, of course, cannot be reared in the dark and they have to be pretty old—five or six months—before they can crawl. When crawling babies were tested, they avoided the deep side of the visual cliff, which suggested that they understood the consequences of crawling off a cliff. But what about infants who are not yet able to crawl? Do they understand what would happen if they were dropped off a cliff?

In the early 1990s, a research team* led by University of California, Berkeley psychologist Joseph J. Campos sought to answer this question.[6] They first tested seven-month-old infants, half of whom had begun to crawl whereas the other half had not. By attaching heart-rate monitors to the infants, the team was able to assess the babies' emotional reaction to the visual cliff. With the babies' parents in the next room, female experimenters lowered the babies over the deep side of the visual cliff. The results: the heart rates of crawling infants went up, indicating arousal ("yikes!") about the chasm that some strange person was lowering them into. But the heart rates of the preambulatory children went down, indicating that they noticed and were *interested* in the cliff but not aroused or frightened. Apparently, the not-yet-crawling babies didn't know what the depth *meant*—they were not physiologically aroused by the possibility of being dropped into the abyss on the deep side of the visual cliff. Why should being able to crawl affect whether babies understood the implications of being dropped off a cliff? To answer this question, Campos and colleagues were inspired by one of the most pro-

---

* The second author, Bennett Bertenthal was a colleague of Denny's at the University of Virginia. Most of what Denny knows about infant development and how to study it, he learned from Bennett.

found studies in all of the developmental literature, a study that showed that *self-produced locomotion*, like crawling or walking, is a crucial part of meaning making.

In the early 1960s, just a couple years after Jackie Gibson's visual cliff experiment, MIT vision pioneer Richard Held conducted a different set of feline studies that would come to be known as the "kitten carousel" experiments.[7] In these studies, pairs of kittens were given visual experience in the experimental phase of the study and otherwise raised in the dark. During the illuminated experimental phase, one kitten had control over its walking, while the other was passively conveyed via a tiny carousel into which both kittens were placed. Imagine a carousel, but instead of horses, there were two baskets, 180 degrees apart, each holding a kitten. One of the baskets had four holes in the bottom through which a kitten's legs could extend to the floor, thereby allowing the walking kitten to drive the carousel. The basket in which the other kitten was placed had no holes, so this kitten was passively taken for a ride. After being raised like this, both kittens could see fine. Their visual systems had matured normally. But the kitten in the passive condition behaved as if it had no idea how to make meaning out of what it was seeing. The passive kitten had problems guiding its paws, it didn't discriminate between the deep and shallow side of the visual cliff, and it failed to blink when something approached its eyes.

Campos and company drew on the kitten carousel study to crack the code for how human babies learn the meaning of heights. As a follow-up to their first experiment, the researchers divided a new set of infants equally into two groups. One group was sent home with "infant walkers" and the other control group was not. Infant walkers are little baby-size vehicles in which a baby can be placed in a seat with its feet in contact with the floor—not unlike the driver's seat in the kitten carousel. The infant walker's seat is surrounded by bumpers on all sides and supported by wheels below. Just like that, a baby who is unable to crawl can now use its feet to zoom around the room. In their experiment, the infants who were provided with walkers had to accomplish at least 32 hours of voluntary locomotion before returning the lab. Then, once again, the heart-rate monitors were applied, and the kids were lowered onto the deep side of the visual cliff. This time, the walker-conditioned infants showed the alarm response of elevated heart rate, while the non-walker infants did not. Even if their mobility was artificially aided by way of the infant walker, the kids' response to the precipice was in line with those who were already naturally crawling on their own.

In tandem, the kitten carousel and the infant walker studies can teach us some key lessons about infant development—and also how, in a more general sense, experience shapes our lives. The first is the matter of *agency:* in order to fully understand what you're experiencing, you need to have had a role in creating the experience. We learn to understand what we see by learning to crawl and then walk. It's like being in the passenger versus the driver's seat of a car. You're much more likely to remember the turns needed to get to a friend's house if you were driving rather than idly riding shotgun. If you were sitting in the passenger seat, you're getting the same visual information, but it's not nearly as meaningful to you and you're not engaging with it nearly as much. Similarly, the kitten in the passive carousel seat received the same information as the active kitten, but for the passively moved kitten, this information could not be related to its own actions. The active kitten created its experience. The passive kitten simply endured it. The same is true for the toddler who can locomote versus the baby who moves only when carried by others.

A second theme in motor development is *enablement:* one thing leads to another. Abilities cascade. Like Campos and his colleagues note in the conclusion of their 1992 paper, "new levels of functioning in one behavioral domain can generate experiences that profoundly affect other developmental domains, including affective, social, cognitive, and sensorimotor ones."[8] A baby learns to crawl—and then walk—and it thereby changes the social ecology of the family. They can explore more of the world if they desire, stay close to the caregiver if they desire, and elicit all sorts of responses from their caregivers as they seek out forbidden and approved objects in the home.

This brings us back to Karen Adolph. Following her revelatory moment with Jimmy's book, she studied with Jackie when she was pursuing her PhD at Emory University in Atlanta. Early in her training, Adolph recalls an occasion when she saw some babies at a day care center climbing onto a cabinet and then not being able to get down. She told Jackie about what she had seen and Jackie replied: "Well, that's interesting, dear. Why don't you follow up on that?" Adolph has been studying what infants and toddlers know about what they can and cannot do ever since.

In 2000, Adolph set out to solve the mystery about why self-produced locomotion experience—crawling or using infant walkers—is a necessary precursor to avoiding or being aroused by the deep side of the visual cliff. Her studies would transform her field. Adolph surmised that using their bodies to locomote taught the kids the virtue of having solid ground under their feet (or hands and knees). Crawling was a way of learning that you could only locomote with firm ground underneath you. You cannot crawl on air.

So Adolph, with the help of her lab's apparatuses, put a group of nine-month-olds to a test. In this experiment, babies were set on one platform, facing a gap, on the other side of which was another platform, on which stood a tantalizing, possibly grab-able toy. These babies were then asked to reach for this toy in two positions: sitting or crawling. The researchers then moved the platform to a range of different distances, noting when the babies made the reach for the toy across from them and when they decided to avoid the danger and stay put.

Babies acquire movement abilities in a set order. First, they learn to sit, then crawl, and then walk. Adolph found that what they learn about distances and the like as a capable sitter does not transfer to when they begin crawling.

Babies who'd become experts at sitting but not crawling were able to estimate when the toy was in reach and when it wasn't with exquisite accuracy
when they were sitting, but they had no clue what they were doing when in a
crawling posture. Almost a third of the babies tested appeared totally oblivious to the risk presented by a too-long gap: "In fact, 6 infants showed finely
tuned avoidance responses in the sitting posture but no capacity to gauge
their ability in the crawling posture," Adolph wrote.[9] "They attempted all
gap distances in the crawling posture, including the 90-cm gap, which was
tantamount to crawling into thin air." Follow-up experiments put this into
starker, startling relief. Faced with a similar gap between platforms, kids who
had become expert crawlers refused to venture across the open space, while
those same children, when they were just learning to walk, would meander
straight off the cliff, like in some particularly sadistic Wile E. Coyote cartoon. Through hundreds of trials concerning different modes of movement,
Adolph and her colleagues repeatedly found that what a baby learns about
distance in one locomotion mode doesn't translate to another. From sitting
to crawling to standing, the kid has to relearn what space means for each
particular form. Indeed, "cruising," the technical term for when toddlers
walk while holding on to furniture in a room, like a beginning ice-skater sup-
porting themselves by holding on to the wall of a rink, doesn't even inform
walking, which, from the observer's standpoint, may appear to be a nearly
identical movement.

When crawling, infants aren't learning some objective truth about what
20 centimeters means in space. Rather, like the kittens in the carousel,
they're learning what those objects and situations mean to them relative to
their bodies and action capabilities. To use the Gibsonian term, babies are
learning the *affordances* of space relative to themselves. That's why the same
gap, cliff, or slope mastered when crawling has to be relearned when walking.
While it might be the same environment, it's a whole new experience for the

kid. When babies learn to crawl, they learn the affordances of crawling—the opportunities and costs that surfaces have for crawling. When they first learn to walk, they behave like kittens raised in the dark. They have no understanding of what the world affords for this newly acquired skill. Childhood is spent learning the new affordances made possible by ongoing maturation, as one way of moving gives way to the next.

Children's understandings of the world are forever playing catch-up with their emerging action capabilities. If you thought puberty was a time of alarming bodily changes, try the ascent into toddlerhood: in the first two years of life, height doubles, body weight almost quadruples, and the circumference of the head expands by a third. And this is not a casual, gradual process but rather one of fits and starts. An infant may grow one to two centimeters in a single night and may shrink by almost a centimeter between when they wake up and when they go to bed. To say childhood is a time of change is quite the understatement.

Every baby is a scientist, and each is constantly experimenting to find the answers to the problems of getting around. Learning to walk reflects "individual, idiosyncratic solutions," Adolph writes.[10] Strategies abound in the first month of walking: "steppers" will take tiny, careful steps to minimize the disturbances to their upright posture; "fallers" tilt forward, asking their feet to constantly catch up to their upper body; while "twisters" swing one leg and then the next like a bored geometry student walking a measurement compass across a desk. Then, after about a month in the ministry of silly walks, kids will start to arrive at the "pendulum mechanism" that marks standard walking. After two months of practice, their walking speed grows faster as their gaits lengthen. And not only do their abilities change, but infants also have to deal with other factors affecting their behavioral abilities. Consider the diaper: Adolph has found that wearing a cloth diaper is "equivalent to losing two months of walking experience" in terms of the maturity of a child's gait pattern, while a thin disposable diaper was equivalent to five weeks. And toddlers will take shorter steps when wearing a diaper and pants compared with just a diaper.

GETTING A HANDLE ON THINGS

YOU MAY HAVE NOTICED that babies are constantly mouthing, throwing, and squeezing things: that's because they're literally discovering how objects in the world work, like so many tiny Archimedeses stepping a foot into a bath and noticing the water level rise. Jean Piaget, the Swiss grandfather of developmental psychology, boldly and presciently said that reality is created in the infant's mind, one gummed toy at a time. (Hence his book title: *The Construction of Reality in the Child*.) By four months, babies can reach and grasp at things, and the empirical stakes get higher: the grabbing, banging, and putting seemingly everything in the mouth are ways that babies come to understand how gravity, three-dimensionality, and the rest of their world function.

During that first year of life, babies discover that when they do a thing, the world responds. (Agency!) Piaget himself observed how delighted his infant daughter became when he tied a ribbon to her foot and attached it to the mobile above her—little Lucienne would laugh, smile, and babble. With the kick of a foot, she could see her mighty impact upon the world. As successive experimenters have noticed when studying this "mobile foot-kicking paradigm," the tot tends to become somber, fussing and crying should the ribbon—and their amplified agency—be taken away. Piaget and the experimental psychologists to follow him would findthat this agency is crucial to making sense of the world. Indeed, as was found in formative experiments in the 1960s onward, infants as young as eight weeks, with a ribbon tied to their foot, will learn the association between kicking their foot and making the mobile move, leading to more kicks over time. From the cradle, we're learning cause and effect, especially if we're helped out with a little bit of tot-friendly technology.

At two or three months of age, babies can swipe or swat at objects in their vicinity, but they can't really grab them. Observing this, Amy Needham, then a researcher at Duke, and her colleagues fabricated what have come to be known as "sticky mittens," infant-size mittens with Velcro stitched to their side.[11] Drawing inspiration from the infant walker study, Needham and her team had an idea. What if, with a little bit of equipment, babies that couldn't grab toys suddenly could? How would that affect their behavior?

Using data from the Durham County, North Carolina, vital records office, the researchers contacted the parents of what ended up being 32 participating infants, half of whom were placed in the experimental group of the study. For this select group, the research team gave parents the sticky mittens, a journal to record their sessions, and a special set of toys with a strip of Velcro attached that mated with the Velcro on the sticky mittens. The parents were asked to show the kids how the sticky mittens worked and to help them play with the toys. With the aid of sticky mittens, the infants could now reach out, swipe at the toy, and—voila!—grab it. The control group was treated exactly the same except that they were not given sticky mittens.

In subsequent testing with familiar and novel toys, the "sticky mittens" infants were prodigious compared to the control group. They spent twice the time visually exploring objects as the control group, and made nearly double the number of swats at the toys. And they were three times more likely to switch between "oral exploration" (mouthing at new objects) and visual exploration (looking at them) compared to the control group.

In a follow-up study that doubled down on the kitten carousel likeness, Needham and her colleagues separated infants into an "active training" (where they had the sticky mittens) and "passive training" (where toys were touched to the babies' hands) groups. Compared with the passive group, the active, mittened infants showed much greater grasping activity—having tasted the joys of grabbing things, they continued to do so even without their sticky mittens. She then followed up one year later to check in on how the kids' exploratory tendencies were shaping up. A year later, the actively trained kids showed greater visual interest in toys presented to them, were less distracted during playtime, and had longer bouts of grasping and rotating objects than the passively trained kids or a wholly untrained control group. This, then, is another cascade. Early motor experience unfurls into greater motor skills.

But reaching, of course, doesn't happen on its own. Karen Adolph has found that posture shapes reaching and grasping and, with that, early cognitive skills. Maintaining an upright sitting posture requires maturation and the development of a highly organized balancing act to avoid succumbing to gravity's pull. She and her colleagues have found that once infants are able to sit up they become much more engaged in manipulating and exploring objects with their hands. Again, one development outcome leads to another. Sitting up frees the hands. Otherwise, it's hard to hold and visually inspect objects while lying prone. Relatedly, Adolph and colleagues have found that babies who are able to sit on their own are also better at recognizing the three-dimensional shape of objects compared to babies of similar age who have not yet mastered sitting. Sitting promotes holding, which leads to exploring held objects and to the discovery of how things look from different perspectives.

DEVELOPING IN ADULTHOOD

DISCOVERING NEW POSSIBILITIES FOR acting in the world—and thus experiencing it differently—does not stop in childhood but stays subtly flexible in adults. Drake encountered a world-class example of that when he sat down to interview NBA superstar Stephen Curry, the Golden State Warriors point guard, six-time All-Star, three-time NBA champion, and twice the Most Valuable Player in the league. Under the direction of his trainer, Brandon Payne, Curry spends much of his off-seasons practicing different sorts of neurocognitive drills. These might involve FitLight Trainers, devices invented by a Danish handball coach who wanted to train his goalkeeper to be more responsive. FitLights are pressable lights, about the width of an adult's hand, that can be stuck to a wall or floor and rearranged endlessly. They can also turn many colors, prompting different reactions. All the while, they can track the trainee's performance, gathering data on how many times they miss or hit the light-up buttons as the prompts come on and how quickly the athletes are reacting. During a training session, different sequences of colors represent different game situations, to which Curry is to respond with different sorts of dribble moves, like a crossover dribble, quickly bouncing the ball from one hand to the next. The purpose is overload, pushing not just Curry's reaction time or coordination but his ability to rapidly learn, read situations, and react. "In a game, there are so many different variables that are thrown at you—the defense, where your teammates are, how fast your body's moving—and you have to be in control of all those decisions. So we overload in our workouts so that the game slows down in real life. It helps you become a smarter basketball player," the MVP said. "I think my ball handling has become a lot crisper, my decision-making has become a lot better, and I feel more creative on the floor. I feel like the game is definitely slowing down, so I can make better moves and have more control over my space out there." Sports science has corroborated Curry's lived experience: the best athletes are elite visual learners. The title of a 2013 *Nature: Scientific Reports* paper about summed it up: "Professional Athletes Have Extraordinary Skills for Rapidly Learning Complex and Neutral Dynamic Visual Scenes."[12]

Curry's observation that time seems to slow down when he is playing well is common among athletes. In describing what it is like to be "in the zone," Bill Russell, one of the greatest basketball players of all time said, "At that special level all sorts of odd things happened. . . . It was almost as if we were playing in slow motion."[13] Similarly, the brilliant professional tennis player John McEnroe described episodes of peak performance as "Things slow down, the ball seems a lot bigger and you feel like you have more time."[14]

Inspired by such comments and her own experience as an elite athlete—she was a gold medalist at the 2005 World Games in the sport of Ultimate Frisbee—Jessi Witt decided to test whether the perception of time is, in fact, influenced by the ease of performance. Recall from the introduction, that it was Jessi who showed that softball batters saw the ball as bigger when they were hitting well. Jessi's first assessed tennis players on their perception of tennis ball speeds.[15] Participants stood behind the baseline and attempted to return balls served by a machine at varying speeds. Following each return, participants pressed the space bar on a computer keyboard for a duration that they felt matched the time that the previous ball had been in the air. Jessi also recorded whether the participant hit the ball successfully into the opponent's court or hit the ball out of bounds. She found that the ball's speed was judged to be slower when the return was successful than when it was not. She replicated this finding using a computer game similar to Pong, an early arcade video game in which the player attempts to block a moving ball with a paddle. Jessi varied the ease of blocking the ball by manipulating the size of the paddle—a tiny paddle made it hard to successfully block the ball and vice versa. Again, she found the ball's perceived speed was related to ease of intercepting and blocking it. Balls seemed to go slower when they were easily blocked with a big paddle as opposed to a small one. In an especially elegant study, Rob Gray, associate professor of Human Systems Engineering at Arizona State University, assessed varsity college baseball players and found that they, too, perceived the size of baseballs to be greater and perceived their speed to be slower when they were easily able to hit the ball due to its location over the plate.[16] When the ball came right down the middle, it looked like a nice, fat pitch to hit.

As our bodies develop and our motor skills are honed, we come to see new affordances in our surroundings. Development, practice, and discovery go hand and hand. With development and practice, the world of gum-able things, becomes one also filled with graspable, throw-able, and perhaps even three-point shootable things. The human *Umwelt* is shaped by what we can do and how well we can do them. Being bipedal endurance animals, no activity is more central to the human way of life than walking.

2.

WALKING

I N 1989, Denny was invited to a three-week workshop at the NASA Ames Research Center in Mountain View, California, a town in Silicon Valley that's now best known as the home of Google's main campus. NASA had gathered some of the world's best vision scientists to work on perceptual problems encountered by helicopter pilots, like perceiving how fast they were flying and how high above the ground they were. Unlike that of birds, the human visual system assumes that our feet are always firmly planted on the ground, and thus, when computing speed, it employs the simplifying assumption that we are always traveling at an unchanging altitude corresponding to our eyes' height.* This assumption works fine if you are in fact walking on the ground, but it doesn't work at all when your altitude changes. As your altitude increases, the ground below you appears to move more slowly, and you feel that you're slowing down. You may have noticed that, when looking out an airplane window, you do not feel that you are hurtling through the air at more than 500 miles per hour. Neither do helicopter pilots. They need to learn how to interpret their speed as their altitude changes. How pilots make this adaptation was the sort of problem that the NASA workshop focused on. Despite the rocket science setting, common sense presided. The workshop scientists took for granted a conventional view of visual processing: your eyes take in visual information, your brain processes it, and the world's structure is accurately perceived. A key assumption in this formulation is that perceptual experience is, for the most part, objectively accurate. For three weeks, Denny tried to persuade his workshop colleagues to consider an alternative assumption—that the goal of perception isn't to give you a geometrically true read of the environment but rather to pragmatically guide the way you think and act.

Mountain View is about an hour's drive from San Francisco, so Denny and his family made frequent visits to the city. There he was confronted with a puzzling fact. San Francisco is famous for its incredibly steep hills. With an incline of about 18 degrees, the section of Filbert Street between Hyde and Leavenworth is most often named as the steepest of San Francisco's streets,

---

* Whenever you move, everything surrounding you appears to move in the opposite direction. This is called optic flow. Take a step forward, and your surroundings appear to move front to back. The speed with which a feature in the environment appears to move is a function of how fast you are moving and how far away it is. Near objects recede quickly, whereas far objects appear to move more slowly. People are known to use optic flow to perceive how fast they are going. To calculate one's speed from optic flow, the visual system needs to know how high above the ground your eyes are. Our visual system assumes that you are walking (or running) and relates the optic flow emanating from the ground to the standing altitude of your eye. This worked great for most of our evolutionary history because locomotion was almost always done via walking. It does not work when flying helicopters.

although for short extents other streets slightly surpass this incline. When viewing streets like Filbert, people estimate the slants to be 50 or 60 degrees, and they will absolutely not believe you when you tell them that the inclines are less than 20 degrees.

Among visual scientists, it was well-known in 1989 that San Francisco's streets are not as steep as they appear. Denny pointed this fact out to his colleagues, using it as evidence for his claim that perceptions need not be geometrically accurate. As far as he could tell, his arguments fell on deaf ears. It was not that Denny's colleagues didn't believe him, but rather that they thought the overestimation of hill slant was just an odd visual illusion. Visual illusions are abundant, but for many visual scientists they are not thought to be very interesting or representative of everyday perceptions. The scientists at the NASA workshop were focused on studying how pilots successfully flew helicopters and they assumed accurate perception as a necessary precondition for doing this well—or at all.

NASA had arranged for all the scientists and their families to live in the same small hotel: a two-story, California-style building encircling a swimming pool. Every evening the families would gather around the pool, play with the children, drink a beer or two; and over the course of three weeks they grew lasting bonds of respect and friendship. Once such bonds are established, it becomes okay for a friend to tell you that what you've just said is "the dumbest thing that they've ever heard in their life." Smiles are shared. Shoulders are shrugged. No hurt feelings. As Denny persisted with his argument that people do not perceive the world with geometrical accuracy, he encountered lots of smiles, shrugged shoulders, but no converts to his position.

Undaunted by the dismissal of his colleagues and sensing that he might be on to something important, Denny looked through the research literature to see if anyone had documented the disconnect between the actual incline of hills and their perceived slant. As testament to visual science's abiding concern with how people perceive the environment accurately, there was just one qualitative study of this perceptual anomaly, in which respondents reported that, yes, slopes did look steeper than they really were.

Denny decided to document this odd perceptual occurrence with a bit more certitude. Back in Charlottesville, he conducted a number of field experiments with his graduate student, Mukul Bhalla,* asking people to estimate the slant of hills around UVA's rolling grounds. Study participants were asked to judge the perceived slant of hills in two ways. The participants and a research assistant stood at the base of the hill in question, where the research assistant asked participants to say aloud what angle they thought the slant of the hill was. The second assessment of perceived slant was a visual-matching task, where participants adjusted the angle of a pie-shaped wedge segment to match the cross section of the observed hill. Participants were also asked to perform an action directed at the hill's incline. Without looking at the device, they adjusted a waist-high tilting board with their hand, so that its surface would feel parallel to the hill.

In experiment after experiment, participants kept overestimating the steepness of hills on the verbal and visual-matching tasks, which assessed conscious perceptions of what the hill looked like.[1] These overestimations were huge! A typical participant looking at a 5-degree hill would judge its incline to be about 20 degrees on the verbal report and visual-matching tasks. On the other hand, the action measure, making the tilt board parallel to the hill, was accurate. So, even though people perceived hills to be much steeper than they were, their visually guided action of adjusting the tilt board was accurate. These findings left Denny totally flummoxed. How can people see

---

* Mukul is now university dean of General Education, at American InterContinental University.

a 5-degree hill as appearing to be 20 degrees and not fall on their face when attempting to walk up it? Given the ubiquitous overestimations of hill slant, successfully walking up hills could not rely on perceptions being geometrical accurate. What, then, is the relationship between our perceptions of the world and the world as it is? How can actions be successful if our perceptions are biased? Without answers to these questions, Denny knew that his scientific colleagues would continue to view the misperception of hill slant as just another odd visual illusion of little importance.

One day as the Charlottesville studies were ongoing, Mukul stopped by Denny's office with a puzzling report. In a recent day of testing, her study participants hadn't overestimated the hills nearly as much as had previous participants. Denny asked Mukul to double-check that the data had been entered correctly. It had. Denny then asked Mukul to look at the original data collection sheets and see if there was anything different about the participants who had been tested on that day. Mukul found that all of the participants were women. With a bit more sleuthing, she determined that all of these women were friends and members of UVA's varsity soccer team. These study participants were highly conditioned Division I athletes. Collegiate soccer players are accustomed to running 7 miles during a game, with midfielders covering a whopping 9.5 miles in a standard 90-minute match. So, it had to be that these women were at elite fitness levels compared to the norm. Denny and Mukul began to wonder: Could fitness, or more generally how easy or difficult it is to physically ascend a hill, shape the way that you see it? To borrow from the Gibsons, could it be that we're seeing the *affordance*—the "walkability" of a hill—when we look upon it?

With this finding rattling around in his head, Denny found himself back in the Bay Area for another NASA research collaboration some months later. He had brought with him a clinometer, an instrument used to measure hill slants, and spent a day walking about San Francisco, measuring the inclines of its steepest hills. As he was measuring an especially steep street, he observed

39

a boy, of about eight years of age, who was helping (actually pushing) an elderly woman up the adjacent sidewalk. The boy was using all of his might in trying to aid what Denny guessed was his infirm grandmother up the steep sidewalk. It seemed to Denny that both of these people could not possibly be having the same perceptual experience of the hill. How could they—for the elderly woman, the hill was too steep to ascend unaided, whereas for the boy, the steepness of the hill would be inconsequential were it not for having to help his grandmother. Could it be that the boy was akin to those soccer players? More generally, if ascending a hill is easy for you, then does it look less steep than it would to someone else for whom climbing the hill would take more effort?

Soon thereafter, Denny and Mukul started following up on their intuition that perceived hill slant was influenced by ease of walkability.[2] A new sample of participants was tested for their slant perceptions while they wore a backpack weighing between a fifth and sixth of their body weight, such that a 110-pound person would don a 20-pound pack, while a 200-pound subject would put on a 35-pound bag. The bags were filled with free weights, allowing the heft to be adjusted depending on the participants, who were either in an introduction to psychology class or were walking by the experiment as it was going on. A control group of participants who didn't have to wear the backpacks also made the estimates. The result: the backpackers consistently saw hills to be steeper than their unencumbered peers.

In another experiment, the same hill-judgment tasks were given to avid runners, who were selected on the criteria that they went out for runs of at least three miles at least three times a week. For this experiment, they were given their start and ending locations, which were both at the foot of hills. Before taking off, the participants made an initial set of estimates of the slant of the first hill, and then set off on long, exhausting runs, lasting anywhere between 45 and 75 minutes. After their runs, they were met by a research assistant at their finishing hill and completed the same battery of slant-perception tasks. Following the exhausting runs, the overestimations increased—up to 45 percent! Another experiment, following up on Denny's earlier discovery, specifically recruited athletes from the cross-country or track team as well as nonathletes. And as predicted the more physically fit someone was, the less their overestimation of hill slant. In another experiment, the participants were older people recruited from a local senior center, with a median age of about 73. They filled out a report on their physical health, and did the same hill-slant-estimation exercises. Consistent with previous findings, the older the participant and the worse their physical health, the steeper the hills looked. Altogether, the results indicate that the slant of a hill is perceived in relation to the capabilities of the perceiver's body at a given time. Denny and Mukul concluded: "Not only is conscious awareness of slant exaggerated, but it is also malleable in that it is affected by people's physiological potential: Hills look steeper to people who are encumbered, tired, of low fitness, elderly, or in declining health. Thus, changes in physiological potential influence apparent slant regardless of whether the changes are short-term and temporary, such as those seen for the backpackers and the joggers, or more long-term or permanent, such as those seen for the fitness and elderly participants. Any change in the capacity to traverse hills will also bring about a change in the conscious awareness of their slant."[3] Put another way: our *walking ability* shapes the apparent *walkability* of the hill, which determines how we see it. You do not see the hill as it is but rather as it is seen by you.

THE HUMAN WORLD OF WALKING

ON A SUNDAY MORNING in November 1974, Donald Johanson and his colleagues Yves Coppens and Maurice Taieb took their Land Rover out for a day of mapping, surveying, and fossil foraging in Afar, a hot, arid region of eastern Ethiopia. It was their second field research season in the area; just a year previous Johanson had found a knee joint, but the team couldn't tell which species of hominin it belonged to. On this round they were after a more impressive, complete find. The stakes were rather large, as the strata that the researchers were working in were older than basically anything else found in East Africa.

At the request of his graduate student, Johanson and the crew went out to the previous day's site in order to precisely mark where they were on the map. Then they put their eyes to the ground to seek out fossils. Just over his right shoulder, Johanson spotted what looked like a perfectly preserved elbow end of the right proximal ulna, the forearm bone that runs from the elbow to the wrist. And since it didn't have the flare on the back that marks monkey elbows, it couldn't be a part of the baboon or colobus fossils that had turned up in the region. This was a hominin, a direct ancestor of humans. But the ulna wasn't sitting on its own. Then came shards of a skull, and up the slope, a whole set of bones just glistening in the sun: femur, ribs, pelvis, and lower jaw. Driving back into camp, that grad student honked the horn and shouted, "Don found the whole damn thing!"[4]

This, clearly, was a big deal: everything else in that geological stratum was more than three million years old, like the pigs and elephants that had already been identified. This new find hugely expanded the fossil record for the whole field. Up until then, "[A]ll of the human ancestor fossils older than 3 million could fit in the palm of your hand, and none were diagnostic enough to be able to say what species they were."[5] While paleoanthropology is usually a game of inferring a puzzle from a single piece (a knuckle here, a rib there), these very old bones already presented a coherent whole. Three million years since this creature died, a full 40 percent of the skeleton was still together. The finding called for a celebration that night, where Johanson put on a Beatles cassette tape featuring the psychedelic anthem "Lucy in the Sky with Diamonds." And while nobody could remember who first made the suggestion, by the next morning the skeleton had a name: Lucy. "Immediately," Johanson says, "she became a person."

Lucy, all 3.7 feet of her, would become what is likely the world's most famous fossil, not only serving as a historic scientific finding but also spurring a surge of interest in the field. Four years later, Johanson officially described her, classifying Lucy as a member of a new species: *Australopithecus afarensis*, or "Southern ape from Afar," a shout-out to Lucy's home region, which, at the time that she was wandering about, was a low-lying woodland. (It was the beginning of a decades-long trend. As he told us in our interview, Johanson and his colleagues have found nearly 500 *afarensis* specimens near Hadar, the village in Afar near where Lucy was found.[6]) By the angle of her pelvis, Lucy was clearly an upright walker. And therein lay the fossil-record-rearranging rub: it would be more than a million years before big brains evolved by way of *Homo erectus*. So Lucy, and her soon to be found peers, also represent a breakthrough as far as causality is concerned: long before the human ancestral line was large brained, we got around on two feet. Crucially, we were bipedal.

Biologists and anthropologists have spent a century and a half arguing about *why* humans are bipedal, but there's consensus that our two-legged, locomotory way of life has been crucial to our becoming the sort of species that we are. In 1871's *The Descent of Man and Selection in Relation to Sex*, Darwin conjectured that bipedalism put early humans a step ahead of other apes. "Man alone has become a biped,"[7] Darwin observed, and one could at least partly see how man came to "assume his erect attitude," which is, after all, one of the species "most conspicuous characters." In other words, walking around on two legs is one of the most obvious differentiators between us and other mammals. And, as Harvard biologist Dan Lieberman notes, one of the key themes in evolutionary theory is that of contingency: one thing follows and is made possible by another.[8] Man becomes bipedal; the hands are freed; and suddenly he can wield a club, cook on a fire, or create art. It's the same "if this, then that" principle at work as we saw in the last chapter on development when, for example, becoming able to sit independently frees the hands to explore objects, thereby leading to an improved understanding of their three-dimensional shape. One thing leads the other, often with unseen consequences.

Bipedalism allowed early humans to traverse great distances, and we evolved into *endurance animals*. Bipedal walking is more efficient over long distances than quadrupedal locomotion, as muscles work less to sustain bipedal gait. This advantage allowed humans to evolve into one of the fastest mammals on the planet—with the qualification that it be for distances of more than 20 miles in the middle of a hot sunny day. This is our species' special advantage.

Under selection pressures deriving from prolonged activity in hot environments, our bodies became covered with eccrine sweat glands that secrete a cooling watery perspiration. Unlike most mammals, we sweat all over. Dogs, for example, can sweat only through their paws. They can run all day in the cold but not in the hot sun like us. Fur acts as an insulation when it becomes wet, and under this selection pressure our body hair evolved to become so fine as to be almost invisible. The obvious exceptions to this being the hair on our heads, where it serves as a hat, and mostly out of sight places having tufts of hair, which have an oily sweat important for social communication via smell. Walking and especially running are promoted by our slender upright bodies, long legs, and big buttocks. Importantly, that newfound sweaty cooling system made possible the cooling of a growing brain—and with the scavenging and hunting abilities that world-class endurance provided, the calories to fuel those brains.

Bipedalism itself is extremely rare: the only other mammals that walks regularly on two feet are kangaroos and wallabies. Some mammals—such as other primates, bears, and even dogs—can walk on their hind legs, but they do not do so well or for very long. The advantage of bipedalism is endurance locomotion. Our species walked out of Africa and did not stop until we occupied every habitable niche on our planet. We are walkers, and our *Umwelt*—our personal worlds—are scaled accordingly. We see our walkable surroundings in relation to the benefits and costs of locomotion, which are high. Most of the calories that we expend in a given day—80 percent—are burned by metabolic processes needed just to keep us alive. We have choice only over how we spend the remaining 20 percent. And of this 20 percent, a staggering 89 percent is burned by walking.[9] Obtaining and conserving energy are currencies of survival, and for us humans, walking is our greatest expense. Denny came to realize that this ecological imperative—to efficiently manage our body's energy use—is the raison d'être for the individual differences in slant perception that his research had uncovered. Hills, stairs, distances—all must be scaled by the brain's need to manage the high bioenergetic costs of walking. It is the most calorically expensive thing that most of us elect to do. The energy that we have volitional control over is spent mostly on locomotion, the life-defining cost of putting one foot in front of the other.

Our species exists today in a moment of evolution as one of a multitude of life's experiments in survival and reproduction. Like all living things, humans possess a unique phenotype, a set of bodily characteristics resulting from the interaction between our genes and the environment. Not only has evolution shaped our bodies, but, in turn, our bodies shape our perceptual experience and our minds.

THE PHENOTYPE AS WAY OF LIFE

DENNY AND HIS WIFE, Debbie, are avid hikers. A few years ago, they were waiting for a train in Dublin, Ireland, which would take them to the west coast of Ireland for a two-week walking vacation. Their attire and gear identified them as hillwalkers—that's what they call hikers on the Emerald Isle—and they were approached by a fellow traveler, who was on his way to a northern region of the island for a hiking trip himself. Itineraries and walking experiences were shared, and as he was leaving, the hillwalker gave Denny a card with a quote from the Danish philosopher Søren Kierkegaard:

> Above all, do not lose your desire to walk. Every day, I walk myself into a state of well-being & walk away from every illness. I have walked myself into my best thoughts, and I know of no thought so burdensome that one cannot walk away from it. But by sitting still, & the more one sits still, the closer one comes to feeling ill. Thus if one just keeps on walking, everything will be all right.[10]

It seems that the melancholy, protoexistentialist Dane was on to something. Kierkegaard would spend hours every day stalking around Copenhagen. Although he lived a largely solitary life, his urban treks gave him a kind of boundaried intimacy with his fellow citizens, taking in the same sights, if not thinking the same thoughts. Philosophy and locomotion have long been joined at the hip. Friedrich Nietzsche went out twice a day with his pipe and notebook while he was calling down ideas from Mount Olympus and deconstructing Christian morality. Immanuel Kant, though he barely left his hometown of Königsberg, knew its landscape well from his daily walks. Charles Dickens would plod around London, thinking through scenes, as did Virginia Woolf. Indeed, one of the fathers of Western thought would imbue our language with his propensity for thinking on his feet: the word *peripatetic* is forever linked with Aristotle, who would walk up and down the Lyceum while delivering his lectures, such that someone who followed him would be referred to as a Peripatetic, from the Greek *peripatein,* or "to walk up and down." For centuries, walking has been a way of getting someplace— physically, as well as mentally.

Experimental psychology has, as of late, started to test this historical association between walking and creativity. One recent Stanford University study found that after going for a walk, participants scored better on tests of divergent or creative thinking, and worse at convergent or analytical thinking. Divergent, creative thinking was assessed through the alternate usage test, one of the classic ways of studying creativity. In it, the participant tries to come up with as many different novel uses for regular, everyday objects as possible, like a tire, a button, a newspaper, or a brick. A brick might be a good paperweight, door stop, or skillet trivet. Convergent thinking was tested with a remote association test, in which the participant has to figure out a word that fits with each word in a set of three. You might be given the triad of *cake, cottage,* and *Swiss,* and then have to come up with *cheese* to get it right. Why should the walkers' performance improve on the divergent thinking test but become worse on the convergent thinking one? The researchers didn't isolate a mechanism, but one possible explanation is that walking—and the way your attention has to constantly survey the scene around you—may stimulate the kind of positive, free-associative daydreaming that's a hallmark of creative insight.[11]

Both walking and running have deep links to wellness. Japanese researchers have spent decades building a literature around *shinrin yoku,* or "forest bathing"—a government-sanctioned practice of getting people away from digital screens and into the woods, with resulting reductions in blood pressure and other physiological stress measures within minutes.[12] Similarly, running, when paired with meditation, has been shown to lessen depressive symptoms in people with major depressive disorder, though researchers aren't sure of the mechanism.[13] And as is familiar to every runner, running is remarkably reliable in its power to clear the mind.[14]

Living outside of the endurance phenotype may incur health risks. What happens when endurance animals fail to exercise? They get sick. Some provocative public health research has found that sales of such labor-saving machines as cars and dishwashers can be mapped to expanding waistlines across the United States. (Sitting is the new smoking, or so we hear.) Take commuting by car, something that 85 percent of Americans do every weekday. There're good reasons to believe it's a health risk: one recent study tracked nearly 4,300 adults in Texas from 2000 to 2007, and measured the distance between their home and work addresses. Commuting distance predicted less physical activity and poorer cardiorespiratory fitness and greater body-mass index, waist circumference, and blood pressure. And, of course, it's not just an American issue. A recent Chilean study of 5,000 commuters found that becoming more active—walking or cycling, for instance—was linked to a lesser risk of obesity or diabetes. What's more, a massive study of middle-aged Britons—some 73,000 men and 83,000 women—found that *active* commuters, people who walked or cycled to work, for example, had significantly lower body-mass indexes than car-only commuters. To zoom out in time scales a bit, obesity can be viewed, at least in part, as being due to an ecological anachronism: we evolved to be endurance animals, but now few people in the developed world have occasion to exercise unless they seek it out in their free time and make it a part of their daily routine. Four or five millions of years of evolution have given us a robust body in anticipation of a life of sustained physical effort that, in today's world, is rarely realized.[15]

We are the only great ape that evolved into an endurance animal. A chimpanzee could no more run a marathon than it could fly. In fact, orangutans, gorillas, chimpanzees, and bonobos—our closest relatives—live quite sedentary lives, spending much of their day lounging about and sleeping. But they do not suffer from the human scourges of inactivity: obesity and diabetes. Why? Their bodies evolved to be sedentary; ours evolved to exercise.[16] Exercise is not the cure for obesity and diabetes; rather, a lack of exercise is a cause of these ailments.

Comparing our walking to how our evolutionary cousins get around can help illustrate how we became endurance animals. Human bipedalism is awesomely efficient. Knuckle walking—the put-your-fists-on-the-ground kind of movement favored by chimps and gorillas—is shockingly speedy, but it is also about 75 percent more bioenergetically expensive than quadrupedalism or bipedalism. It's important to make comparisons to chimps because, according to recent molecular data, they and bonobos are the apes that we've most recently diverged from, with our lineages parting ways around six to ten million years ago. Like orangutans and monkeys, chimps are well adapted to a fruit-eating, branch-swinging way of life. Daniel Lieberman notes that knuckle walking was a kind of reevolution—a mode of ad hoc quadrupedalism that enabled chimps to preserve features of hand, wrist, and shoulder, which allowed them to swing through the treetops, while also being able to mosey along forest floors. That knuckle walking isn't calorically efficient makes little difference to chimps, since they rarely stray more than two to three kilometers away from their home nests. The fossil evidence suggests that the hominins that would give rise to humans found bipedalism to be adaptive, because their environment was changing. While it's impossible to verify with absolute certainty, the leading hypothesis for why bipedalism evolved four to five million years ago is that upright walking was a way of reacting to climate change and resulting changes to habitat. If drought causes the local sources of drinking water to dry up, then an efficient endurance animal can walk away in search of a more favorable home.

Bipedal walking is a response to the pressures to *fit* into a particular environmental situation. Once our human ancestors began to reliably locomote in this way, selection pressures to optimize the energy expenditure required for long-distance walking came into play. With *Homo erectus,* we see a body built to walk and run for extended periods of time in hot conditions. Gathering fruits and vegetables, scavenging, and persistence hunting became ways of obtaining food. In persistence hunting, a group of people will track and chase an ungulate—hooved animals like antelopes—for hours until it becomes exhausted and can be killed with a pointed stick, which was the only weapon *Homo erectus* possessed. Persistence hunting is still practiced in Africa today.[17] *Homo erectus* spent most of his awake hours walking and running. We've inherited his body.

For those who love to exercise their endurance bodies, there is an almost limitless offering of running events available across the United States. For example, in 2016, there were about 30,000 races in the United States—ranging in distance from 5 k runs to marathon, ultramarathon, and Ironman races—which is about 10 percent greater than the number of races available in 2012.[18] About 56 million Americans participated in running, jogging, and trail running in 2017, up from about 39 million in 2006.[19] People routinely run for the pleasure it brings, and their numbers are on the rise. Our built-for-endurance bodies are also what allow Nepalese porters to haul loads of up to 183 percent of their own body weight over Himalayan passes for days on end. The secret is the right pace: walk slowly for many hours each day, rest frequently, and carry the heaviest load possible. Walk for about 15 seconds, rest for 45 seconds, and then repeat again and again and again. If the porters walked for longer durations or rested for shorter periods, then they would become exhausted and could not continue.[20] If they were more conservative and walked less and/or rested more, then they would not travel as far over the day. Biomechanical analyses showed that this regime is ideal. Our gait reflects eons of natural selection for long-distance locomotion efficiency, which is supported by our moment-to-moment perception of our surroundings in terms of the energy cost of walking.

LINKING PHYSIOLOGY TO PSYCHOLOGY

WHAT DENNY'S EARLY STUDIES revealed was how our perception reflects the relationship between what we are attempting to do and what is available in our surroundings to achieve these goals. Consequentially, we need to reconsider our commonsense notions about vision as being an objective video camera of sorts, sending images to your brain. Rather, vision is there to help guide your actions, which suggests that what we perceive depends on whatever aspect of the body or self is relevant for the task at hand. In the case of perceiving a hill, the energy required to ascend it serves as the perceptual measuring stick. For a cereal box on the top grocery shelf, it may be the vertical extent of your arm's reach. So, the way you perceptually measure an extent isn't with a meter or yardstick—the objective tools of the geometer—but rather with that aspect of your body that is relevant for what you are trying to do.

Colorado State University professor Jessi Witt and her colleagues tested this affordance notion in a real-life situation, pulling community members aside at a local superstore. The 66 participants—who were about evenly split between normal weight, overweight, and obese—were asked to estimate the distances of cones on a path set at 10, 15, 20, and 25 meters away from them. Then they filled out a survey, reporting their height and weight, and also how they evaluated their current weight (as in "too low, a bit low, good, a bit high, or too high.") The results: the more overweight people were, the farther they judged distances to be. And here's the twist: the self-evaluation of heaviness—whether they thought themselves to be overweight or not—had nothing to do with the distance estimations. What mattered was their actual weight relative to their height.[21] This finding is reminiscent of Denny's backpack studies. Like hill inclines, distances are perceptually scaled by how much energy is required to walk across them. Backpacks and excess weight increase this energy cost.

The practical reality of these findings is that as the capacity of your body changes, so will your experience of the world. Frank Eves, a public health psychologist at the University of Birmingham, UK, has followed dieters as they do or don't lose weight, administering slope perception tests along the way.[22] True to form, they perceive slopes as less steep as their body-mass index lowers. Eves has also spent more time than most staking out shopping malls—all in the name of science. He tracked people as they either took the stairs or opted for the escalator. Unsurprisingly, the people who were either overweight or carrying heavy packages tended to take the escalator more often than those who weren't carrying any extra weight. Parallel to that, he also found that people who were encumbered by body mass or physical objects estimated the stairs to be steeper than the people who were not so encumbered.[23] But even with all this psychological research, what was needed to really solidify these findings was direct evidence that the effective variable was physiological differences in people's ability to expend energy.

Investigating this direct link required help from an expert in exercise physiology, UVA professor and chair of the Kinesiology Department, Arthur Weltman. Art was about to conduct a study on the role of carbohydrate ingestion to the blood lactate response to exercise, and Denny's graduate student Jonathan Zadra, now a faculty member at the University of Utah, convinced Art to add some assessments of distance perception to his study. The participants were competitive bicycle racers and they were seen in the lab on four occasions. During all experimental sessions, the athletes rode exercise bicycles while attached to intravenous lines that measured their blood chemistry. In addition, they breathed through a mask, which measured oxygen intake and carbon dioxide exhalation.

In two experimental sessions, the cyclists rode the bikes for 45 minutes at a high level of exertion.[24] During one of these sessions, they were given a carbohydrate-sweetened Gatorade to drink, whereas in the other session they were given Gatorade containing a noncaloric sweetener. (Previous research showed that people could not identify the drinks from their taste.) Neither the cyclists nor the experimenters knew which version of Gatorade they were getting. Perceived distance assessments were made before and after each session. It was found that following the tiring ride, participants perceived distances to be far greater if they had been drinking the noncaloric drink relative to when they were given the caloric one. These results indicate that perceived distances are influenced by our caloric reserves—the more fuel you've got in the tank, the shorter walkable extents appear. Note that this all occurs outside of awareness as the cyclists were not told whether they had received the calorically sweetened drink or not.

The findings resulting from the other two experimental sessions are even more revealing. During these sessions, the cyclists rode until they became exhausted. In essence, they were required to ride the stationary bikes, turning the resistance up repeatedly at predetermined intervals, until they could no longer pedal the bike. The dependent measure of interest was VO_2 max at lactate threshold. VO_2 max is the amount of oxygen that a person can inhale during exercise, and it's a standard measure of fitness within sports science. When exercise becomes so demanding that the available oxygen is insufficient to burn the calories needed, then muscles burn calories without oxygen—anaerobic exercise—and lactate is released into the blood. VO_2 max at lactate threshold is the maximum amount of oxygen that a person can inhale at the moment when they hit the boundary for anaerobic exercise and lactate begins to appear in the bloodstream, making it the gold standard for measuring physical fitness. It was found that VO_2 max at lactate threshold predicted the cyclists' distance perceptions. The more fit the cyclist, the shorter the distance they perceived extents to be. This was even true for the assessments of perceived distance made before the cyclists began riding the bikes during experimental sessions, meaning that when the cyclists first walked into the lab, their physical fitness predicted their distance perceptions. This study shows that extents are measured by the amount of energy required to traverse them, which is a function of individual fitness.

Twenty years later, Denny had finally nailed down why the women on the UVA soccer team perceived hills to be less steep than the nonathletes who were typically tested. The greater one's fitness, the more economical is locomotion, and thus the lower the energy costs incurred when traversing walkable extents. Since perceived distances are measured by the energy costs associated with walking, the greater one's physical fitness the shorter the perceived extents to be traversed.

Consider what is meant by the term "fitness." As Denny and his coauthors note, with greater fitness you get greater energy storage in the muscles, greater movement efficiency, and faster and more efficient metabolism. The muscles get packed full of more mitochondria, enabling endurance. "It is these bioenergetic consequences of what is termed 'fitness' that allow athletes to run farther, faster, and longer, or generally outperform less-fit individuals in a physical activity," they observed. "They have more stored energy available, they use less energy to do the same things, and they get more energy and can produce more biological work out of the food they eat."[25] That's the stunning thing about all this research: it shows that each of us lives in our own personal *Umwelt,* a perceptual world scaled by our unique abilities. Moreover, these personal *Umwelt* are dynamic. It's borderline trite to say that committing to an exercise routine can change the way you see things. But the reality is, it can literally change the way you see the world.

Physical *fitness* affects how *you* fit in the world. Common sense suggests that we see the world as it is, but, rather, what we see is how we fit in the world. To paraphrase the ancient Greek philosopher Protagoras, the body is the measure of all things. This is the lesson of the streets of San Francisco.

3.

GRASPING

To someone holding a hammer, everything looks like a nail.

O R SO GOES THE APHORISM that was rattling around Jessi Witt's head in 2008. The saying was something that was often in the air when she was in Denny's lab at UVA, and she carried it with her to her first faculty job at Purdue University in Indiana. "If it's true that holding a hammer makes everything look like a nail," she remembers thinking, "then what about holding a gun?" She was musing about this question with an enterprising undergraduate who, to her delight, took the initiative to put together the experimental materials needed to find out the answer. The student created a set of pictures of a man, wearing a black jacket and ski mask, who was thrusting one of two objects toward the camera, either a gun or a Converse All Star sneaker.

Jessi joined forces with James Brockmole, a social psychologist at the University of Notre Dame, and together they designed a set of complementary experiments in which participants viewed the pictures and for each one indicated as quickly as possible which object was being held, the gun or the sneaker. In four out of five of the experiments, one group of participants held a toy gun whereas the other group held a foam ball. In the final experiment, instead of a gun, one group held a shoe. And the result? If you're holding a gun, then you're more likely to perceive the man as also holding a gun; and, likewise, if holding a shoe, then you are more likely to see him holding a shoe.[1]

This research demonstrates that what we're doing with our bodies—specifically our hands—shapes the world that we experience. Our personal, perceptual worlds are malleable; they change depending upon what we're trying to do and our abilities to pursue these ends. Moreover, these abilities can be augmented by the tools we employ. Weapons represent a particularly worrisome case of tool use. Witt and Brockmole wrote, "It is true that the action-induced biases we observed were not specific to guns. That said, while the bias created by holding a shoe is benign, the act of wielding a firearm raises the likelihood that nonthreatening objects will be perceived as threats." They went on to conclude, "This bias can clearly be horrific for victims of accidental shootings. According to the American Civil Liberties Union, approximately 25 percent of all law enforcement shootings involve unarmed suspects and, although it is impossible to derive a precise number, it is certain that many similar accidental shootings occur among private citizens. It is, therefore, in the public's interest to determine the factors that can lead to accidental shootings as well as measures to reduce the impact of these factors. While several factors including one's beliefs and expectations have been previously identified, the current results indicate that the mere act of wielding a firearm raises the likelihood that nonthreatening objects will be perceived as threats."[2] If we can turn ourselves into reaching, grasping, or even dancing machines, then we're more likely to perceive others as doing the same. More grimly, if we are holding a gun, and thereby have turned ourselves into a gun-shooting machine, the research suggest that we're more likely to see someone, who is actually holding a benign object, as grasping a firearm instead.[3] Rather than a good guy with a gun being the proposed solution to a potentially violent situation, as some firearm advocates argue, it appears that merely holding a gun makes it more likely to see firearm holders around you. How do you see the world? Turns out that, in many cases, it's in your hands.

OUR HANDS—AND ACTIONS— HAVE A MIND OF THEIR OWN

IN MAY 1988, Mel Goodale and David Milner, both young vision researchers at the University of St. Andrews in Scotland, received a call from a colleague at the nearby University of Aberdeen. The news: a young woman living in Italy had suffered a tragic accident. The gas hot water heater in her bathroom was not properly ventilated. She fainted and fell into a coma from carbon monoxide poisoning. This incident left her with an exceptional visual impairment stemming from hypoxia, a condition in which her brain was temporarily starved of oxygen. Their colleague asked whether they might wish to examine her when she returned home to Scotland. They said yes, though when the vision tests came in, things didn't look good. Her visual world was a blurry, uninterpretable, array of colors and blobs. And yet—and here's what is so surprising about this case study—while she couldn't *tell* you that you had a pen in your hand because both your hand and the pen would appear to her as amorphous blobs, if you offered the pen to her, she'd have no problem reaching out and grasping it.

The patient, whose pseudonym became Dee Fletcher, could not recognize her mother when looking at her face but could do so when hearing her voice. If Fletcher was shown a pen, then she could see that there was something there but could not tell whether its orientation was vertical or horizontal. Given her severe visual impairment, it was astonishing that she could walk, navigate terrain, and avoid tripping over things. She could draw simple objects from memory—a boat, a book, an apple—but could not reproduce an illustration when looking at it. She couldn't tell you whether an object was a sphere or a cube, but if you offered it to her, her hands would open to accommodate the object's size, shape, and orientation. When looking at objects, she would just see colored blobs, as if everything had turned into globs of clay, so that a mobile phone and a small shoe would look pretty much the same. She couldn't tell their shape or identify what category of person or thing they belonged to, but she had no problem picking them up or touching them. "To us," Goodale said when speaking to us, "that was a revelation." In long hours in the lab in Scotland and then in Canada, where Goodale would take a job at the University of Western Ontario, Fletcher slowly revealed something profound. How can someone adroitly pick up an object when they are unable to accurately see its size and shape? The answer is—so the evidence indicates—that humans don't have one visual system. They have at least two.[4]

According to Goodale, vision serves two functions: *knowing* and *doing*. Each of these functions is performed by different areas in the brain. In the study of Fletcher and other patients, Goodale and Milner uncovered evidence that the information leaving our eyes flows into dual streams of neural processing. The first is the visual *what* system that provides a conscious awareness of what is in front of you. The *what* system is responsible for our ability to recognize the form and meaning of our visual world, the scene we see before us. The second visual-processing stream is the *how* system, which is responsible for the visual guidance of actions. If you are thirsty, then you may look about for something to drink. It is the *what* system that sees a glass of water, luckily within reach. It is the *how* system that controls your reaching, grasping, and bringing the water to your mouth. Dee

Fletcher had profound damage to her *what* system. She could neither see the form of objects nor identify what they were; her visual world was filled with blobs. Yet her *how* system was intact. She could walk without bumping into things. If thirsty, she could not recognize a glass of water nearby, but if its location was pointed out, then she could pick it up and drink it without any trouble. "She was the Rosetta stone that helped us unpack the story," Goodale says.[5]

To put it technically, the *what* system, according to Goodale, is *allocentric**—it traffics in the world out there, observing the scene you see before you. This vision-for-knowing system is an achievement of the temporal cortical lobe, which is located at the bottom of the brain, near your ear. This is the visual pathway that ultimately results in conscious perceptual experience, your *Umwelt*. It lets you know what you're seeing, perceiving and identifying the shapes around you, giving rise to meanings—that's a chair; that's a cat; that's your partner. As such, to call back to the development chapter, the knowing-what system is agentic: it supports choices about what to do—you see an apple hanging from the tree and decide to grab it. The action system is *egocentric*. Not in the sense of being selfish but rather as being related to the self, not as a passive *observer* but as an *actor*. This *how* system is an achievement of the parietal lobe, which is the top-back region of the brain, near where bald spots first appear. This vision-for-doing is *not* conscious, to the point that some researchers refer to it as a zombie—it has no will of its own. It guides actions without knowing, or needing to know, what it is doing—something that Dee Fletcher illustrated thoroughly. The two visual systems operate in tandem, their cooperation hidden in plain sight—we will an action and our body performs it. Our perceptual world is constructed

---

* *Allocentric* is a combination of the Greek root *allo,* which means "other," and -*centric*. In a social psychological context, being allocentric means caring for other people. And as we'll get into in the last third of the book, *alloparenting* means to care for children that aren't your own.

by the *what* system, which gets to decide what we will do, while leaving the actual control of our actions to the *how* system.

"I think what happened is natural selection has resulted in two rather different but interactive pathways," Goodale explains. "The way we look at it is, we need vision to represent the world to make plans, choose between goals, talk about plans and goals. In order to do that we need some representation of the world that is real and powerful, and vision makes a terrific contribution to that effort. But the computational problems in erecting the world are quite different from the computations our brain uses to guide actions through the world."[6] Just knowing that that glass of water sits on the table next to you isn't enough to get it to your lips. Knowing where it is has to be translated into a sequence of muscular moments that will reach for the glass, grasp it, pick it up, and bring it to your mouth. This is the work of the action system. Opportunities for action are consciously perceived—they make up our *Umwelt*. The visual guidance system is an obedient servant, bringing desired actions to fruition.

Ironically enough, the action system operates without our really noticing. Goodale compares it to the Mars rover. That robotic space explorer is a semiautonomous vehicle by necessity, since the signal transmission time from Earth to Mars is just too great to allow for a real-time remote control as with a drone or a toy car. The human operator down on Earth looks out at the surface of the Red Planet through a camera on the rover, spots something interesting, and tells the robot to go over there and sample the rock. The dutiful rover is like the *how* system waiting for the *what* system's instructions. Its onboard guidance system is capable of making the necessary movements, but it needs to be told where to go and what to do. Same with us. We reach out, and our hand scales to the object just prior to the grasp. "The cognitive system needs that visual representation to do its job, and the motor system needs visual coordinates to get your hands on the glass," Goodale says.[7] Again, the way we see is not purely *optical*, or simply the result of eye and brain in isolation. Our perceptions are scaled to our ability to interact with the world and perform tasks, as Denny and his colleagues' work have shown. These knowing and doing systems have a reciprocal relationship. It is through doing that we come to know about the world. Not only are we locomotion machines, seeing the world in relation to walking affordances, we are also manipulation machines and see the world in terms of manipulative opportunities. Through exploring, we come to know.

The hands are guided by the vision-for-action *how* system, and, as we've seen, they need to accommodate our movements to the world as it is and cannot be seduced by biases inherent in the vision-for-knowing system. In later experiments, Goodale investigated how the *how* system can't be fooled through, among other things, the Ebbinghaus illusion, named for its discoverer, the German psychologist Hermann Ebbinghaus.[8]

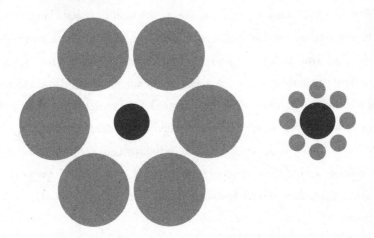

It's clear that the middle circle on the right looks larger than the one on the left. But in fact the two circles are the same size. It tricks the eye, but less so the hand. In one highly cited experiment, Goodale and his collaborators set up a tabletop to present the illusion with thin poker chips of various sizes taking the place of the center circles. The participants saw the illusion, which was reflected in their size estimates, but when asked to pick up that center poker chip, their hands scaled much closer to the chip's actual size—not the inflated or deflated sizes the illusions lead their *what* systems to see. This has a striking parallel to Denny's work. Recall how, in his hill slant-perception studies, he asked participants to make verbal estimates, match the angle seen to the hill's incline, *and* tilt a board so that it felt parallel to the hill's incline. The hand-adjustment task proved to be far more accurate than the verbal or visual estimates. The action system, in other words, was more accurate than the knowing system.

Why might this be? And why would the *knowing* system, the one that reflects commonsensical notions of vision, be so easily fooled, so readily malleable? It goes back to the point Denny was trying to make to his colleagues at the NASA Ames Research Center. That, given the pressures of evolution, the job of conscious perception isn't to give you an objectively accurate view of the world but rather to help you make practical decisions about what to do. In the case of the Ebbinghaus illusion, the middle circle on the left is *smaller* than the circles that surround it. The middle circle on the right is *bigger* than its surrounding circles. In our conscious *what* vision, the middle circles contrast with their surroundings making it easier to see their relative sizes as small or big. In the case of perceiving the slant of hills, historically our species' survival depended on the judicious conservation of energy, so we see slopes scaled to the bioenergetic costs of locomotion. But regardless of that scaling, the *how* system has to deal with what's physically there—hence the greater accuracy.

Goodale's breakthroughs provide empirical evidence for an insight that's metaphorically embedded in the English language. When we *see* the point a friend is making in an argument, we flatly appreciate it, but when we truly get what the person is saying, we *grasp* the ideas he or she is trying to articulate. Knowledge is not only what we possess but also what we can manipulate. Like our feet and mouths, our hands are a primary physical interface with the world. To grasp the human condition, then, we need to investigate what it means to grasp—its evolutionary origin, the way it shapes vision, and how it shapes the way we think.

The hand serves as a guiding light and central interface to our experience, our culture, and our identity. Worldwide, the first visual artists were stenciling their hands against walls, as far back as 39,000 years ago. Given that our hands are the primary mode of manipulation for the body, they are also a primary means of expression, and they also play a key role in how we make emotional sense of the world. Long before Darwin, philosophers would wrestle with the meaning of the hand. Aristotle chided the pre-Socratic philosopher Anaxagoras for thinking "the possession of these hands is the cause of man being of all animals the most intelligent." Instead, Aristotle maintained, "[I]t is rational to suppose that the possession of hands is the consequence rather than the cause of his superior intelligence."[9] But the evidence weighs against Aristotle and the early neurocentrism he articulates: what we know about *Homo erectus* supports the view that humans first evolved dexterous hands and later evolved a brain of sufficient size and complexity to exploit their use. But Aristotle nonetheless greatly admired our upper appendages: in his *De Anima,* or *On the Soul,* he describes the hand as "the instrument of instruments," since it can take up instruments, thus allowing them to *be* tools in the first place.[10]

For Darwin, becoming bipedal was an inflection point in more than one way. It would lead to humanity's world-class ability to endure, especially over long distances and hot days. But it also freed the hands. Not needing to swing from branch to branch or knuckle walk, our forebears had hands available to do other things. Consequently, our hands developed exceptional dexterity, further differentiating the human form from our evolutionary peers. Humans are the only primate that can touch the tip of each finger with the tip of the thumb. This was a game changer for tool use and the suite of behaviors we now call fine motor control—imagine trying to thread a needle without being able to touch the tip of your index finger with the tip of your thumb. Indeed, evolutionary theorists speculate that tool use, which became available through nimble hands, actually laid the groundwork for abstract reasoning: if you can untie a physical knot, then you can untie a knotty conceptual problem. Your hands perform what's called "recursive embedding" whenever you tie your shoes: *first this, then that, and that within this.* According to the linguist Noam Chomsky, recursion is a defining characteristic of language: *John thinks that [insert any statement here] is a laughable idea.* No other animal shows evidence of recursive embedding in its communicative system. And it may owe to the neural control of our complex and flexible hands, with their spectacular ability to manipulate, carry, make, use tools, and touch others.

On his desk at UVA, Denny keeps a trove of manipulanda—things he likes to handle while thinking, like a seashell, a guitar pick, and a conductor's baton. In 2002, he was meeting with Jessi Witt, then a graduate student, when she proposed that having a tool that extended reach ought to shorten the perceived distance to objects that were in reach with the tool but out of reach without it. So they took the baton and pushed a paper clip around a table until a full-blown research design had emerged. The hunch proved true: objects that are out of arm's reach but reachable with a tool appear closer when the tool is wielded than when it is not. Thinking about using a tool is so much easier when the actual tool is being held—the insight is easier to grasp. Yet as the case study of Dee Fletcher demonstrated, grasping itself may not require seeing, at least in the conventional sense.

Mel Goodale was a student of Lawrence Weiskrantz, a Cambridge psychologist who was a pioneer in the study of what would become known as the two visual system. Weiskrantz, a researcher with the wit of a poet, was given to coining terms, and he was the first to observe what he'd call blindsight.[11] As its name implies, blindsight, refers to the ability of the blind to see. Although this sounds nonsensical, it is not—especially looking back from Goodale and his colleagues' work. Clinical researchers have repeatedly found that some people with no conscious visual experience can nevertheless walk down a cluttered corridor without tripping over things. Their visually guided walking allows them to walk around obstacles even though they have no awareness of the obstacles that they are skillfully avoiding. They are blind and have no conscious visual experience, yet, in a pragmatic sense, they can navigate as though they see.

Some of the first studies on this phenomena were conducted by Weiskrantz and his colleagues in the 1970s.[12] One of his best-known case studies involved a patient given the initials D. B. From the age of 14, D. B. experienced severe headaches, which were preceded by flashes of light. By his late 20s, these headaches had become more frequent and pronounced and the light flashes had grown more vivid and lasting. A tumor was discovered on his right primary visual cortex. Following its surgical removal, the headaches were relieved, but D. B. was rendered blind in the left half of his visual field. To get a sense of this blindness, stick your right index finger out in front of you at arm's length, look at the finger, and now imagine that you could see nothing but blackness to the left of the fingertip, whereas your vision to the right was perfectly normal. The area of blindness is called a scotoma. This was the condition of D. B. He felt himself to be completely blind to everything to the left of where he was looking.

Weiskrantz presented D. B. with a number of tasks designed to assess whether he had any visual functioning in his blind visual field. D. B. would be told that there was something in his blind region but was not told what it was. The results were startling. "Even though the patient had no awareness of 'seeing' in his blind field, evidence was obtained that (a) he could reach for visual stimuli with considerable accuracy; (b) could differentiate the orientation of a vertical line from a horizontal or diagonal line; (c) could differentiate the letters 'X' and 'O.'"[13] Even though these feats were accomplished with accuracy, D. B. had no visual awareness of what was presented in his blind field. Weiskrantz wrote of D. B., "But always he was at a loss for words to describe any conscious perception, and repeatedly stressed that he saw nothing at all in the sense of 'seeing,' and that he was merely guessing."[14]

Numerous studies have confirmed Weiskrantz's findings. Early on, there was some controversy about what underlying neural mechanisms were responsible for blindsight. Weiskrantz's position was that blindsight was achieved via subcortical visual pathways that bypassed the visual cortex.

Support for his position comes from numerous investigations including the case of T. N., a man without any visual cortex. T. N. was another patient studied by Weiskrantz and colleagues.[15] T. N. had suffered two massive strokes that destroyed both his right and left visual cortices. He was rendered completely blind; neuroimaging studies showed that he had no functioning in his primary visual cortex. T. N. walked with a probing, obstacle-avoidance cane and required guidance from a sighted person when walking in unfamiliar environs. Yet, once set on a path, he could navigate a space skillfully: experimenters once filled a corridor with boxes and other clutter, randomly placed about the floor. T. N.'s cane was taken from him and he was asked to walk down the corridor. He did so skillfully without once tripping over or bumping into any of the obstacles and without the aid of any conscious visual experience. He stepped around objects even though he had no awareness of the obstacles in his path.

Not everyone who is blind is capable of blindsight. If blindness is caused by damage to the eyes, then no visual functioning is possible. Blindsight is the result of damage to the visual cortex, through which visual information must pass if it is to become conscious. However, other visual pathways exist that bypass the visual cortex, and these pathways support visual functioning without visual awareness. Common sense suggests that we first experience the world and then act on it. But in many situations, visually guided actions can be successfully performed without conscious visual experience. This is the lesson of blindsight. Our perceptual world is for making choices about what to do, not for controlling the doing itself. We possess two systems: one a decider and the other a servile doer.

HANDS GUIDE ATTENTION

TRY THIS SIMPLE AND easy demonstration. Extend your arm and look at your index finger. Now move your finger left and right and track it with your eyes without moving your head. Easy-peasy! This feat is called a smooth-pursuit eye movement; and among all the mammals on Earth, it can only be accomplished by primates.[16] If you have a dog nearby, then try this: hold a dog treat in front of the dog and move it left and right. The dog will keep its eyes fixed on the treat, but will do so by moving its head, not its eyes. For a dog, smooth-pursuit eye movements are impossible. Now, it is not the case that dogs and other mammals can't move their eyes. They can but only with fast, jerky eye movements called saccades, which we make as well. Saccades are for scanning our environment to see what's there. Smooth-pursuit eye movements are for tracking things of interest, especially things that are held in our hands. Threading a needle requires smooth-pursuit eye movements that follow your fingertips, the needle, and the end of the thread.

Evolutionarily speaking, the dexterity that developed in our hands created opportunities for manipulation that could only be fully realized by acquiring greater dexterity in our eyes. As a consequence, only those animals with hands—primates—are able to make smooth-pursuit eye movements. This is a wonderful example of how abilities cascade—one thing leads to another. Having hands created a selection pressure for having eyes that can track their movements. But the cascade does not stop here.

The hands define "regions of space upon which the spotlight of attention should shine most brightly," as one set of researchers recently observed.[17]

Consistent with this notion, targets near the hands can be detected more readily when they are easy to grasp.[18] Viewing objects near your hands also makes you more detail-oriented and better able to ignore distractions.[19]

One of Denny's graduate students, Sally Linkenauger, now a senior lecturer at Lancaster University, UK, wondered how potent the hand's scaling power could be. She had research participants look at graspable objects, like baseballs or Ping-Pong balls, through magnifying goggles, which made these familiar objects appear huge. Participants then put their dominant hand near the objects. Not only did their hand look normal in size, but the balls suddenly shrunk back to their actual size when viewed next to the hand. The effect is quite startling as the balls appear to change size right before your eyes.[20]

Linkenauger has also found that right-handers estimate their right hands and arms to be bigger and longer than their left hands and arms, while left-handers perceive them to be of equal size, perhaps owing to their living in a right-hander's world. Similarly, if a righty is looking to reach for an object with their right hand, it will appear closer to them than if they wanted to reach with their left hand. It's once again affordances: when you turn yourself into a reacher, you experience your perceptual world in terms of your reaching abilities, which for most of us are heavily biased by right-handed dominance.[21]

ON THE OTHER HAND: LATERALITY AND SEEING, FEELING, THINKING

YET ALL HANDS ARE not created equal. Experientially speaking, the vast majority of us heavily favor one hand over the other. Our spines may mark a physical midline, but about 90 percent of the human population interacts with the world primarily through their *right* hands, and researchers estimate that's been the case for 5,000 years—though nobody really knows why.[22]

University of Auckland scholar Michael Corballis describes this heavy right-handedness as "the apparent contradiction between structural symmetry and functional asymmetry."[23]

The hand doesn't just guide how we grasp objects, it also shapes our ideas and feelings. One of the most inventive researchers in this domain came to it offhandedly—Daniel Casasanto, now a psychologist at Cornell, who started his empirical adventures under Steven Pinker at Harvard. Under Pinker's guidance, Casasanto was pursuing a popular idea in linguistics, championed by Noam Chomsky: the hypothesis that cognition, deep down, is the same for everyone, with different languages being like clothing that dresses it in one garb or another. But the data, stubbornly enough, proved otherwise, and that meant changing his adviser and beginning a new research program. So Casasanto began studying, among other things, "linguistic relativity," which basically says that the language you use shapes the way you think or perceive in some significant way, either individually or as a culture— like how cold cultures have lots of words for snow.[24] Recent experimental research has found more specific, interesting evidence of the theory. One study found that people who are native speakers of Russian, which doesn't have a single word for *blue* but rather separate words for light blue and dark blue, were 10 percent faster in identifying different shades of blue than native English speakers.[25] Over the years, Casasanto would give talks on linguistic relativity at academic conferences and the like. He kept getting one question: What's so special about language? Why would language, of all things, have such a special influence on our conceptual lives? Casasanto came to believe that language recruits everyday cognitive forces and organizes them in certain ways, pushing us to pay attention or remember things in different ways. But it wasn't just the words we use that could be affecting how we think and feel, he came to realize, but also the bodies we have and how we use them.

He had been intrigued by—and skeptical of—the arguments of George Lakoff and Mark Johnson. This linguist-philosopher duo, respectively of

California-Berkeley and Oregon universities, have made the trenchant argument that for language to mean anything, it has to be grounded in direct, bodily experience. In the influential *Metaphors We Live By* and other works, they make the case by way of observation and thought experiment, tracing the many odd ways that physical actions and perceptions show up in our everyday expressions. Like how time is "long" or "short," though it has no literal distal extent; that numbers are "big" or "small," though they have no literal size; or that it's all "uphill" or "downhill" from here, depending on who you ask. Casasanto wanted to test Lakoff and Johnson's arguments experimentally and assess whether people *thought* in these embodied metaphors. To do so, he'd need to take language out of the picture.

It began with marbles. He and his colleague Katinka Dijkstra, of Erasmus University in the Netherlands, placed two boxes of marbles, one stationed on a higher platform, one on a lower, in front of participants. In some trials, they asked participants to move marbles up from the lower to the higher box; in others, from the higher to the lower. While this vertical action was going on, the researchers asked participants to tell simple stories about themselves, like a memorable birthday, what they did last summer. The people idly moving the marbles up tended to tell positive autobiographical tales, and those moving them down trended toward stories of bad luck and missed connections. Without their realizing it, the upward or downward movements were guiding the emotional tone of the stories they told—either on the upswing or the down.

To Casasanto, this was a clue: "If you activate these mental metaphors," he says, doing so "has causal power."[26] In a follow-up 2019 study, Casasanto put this finding into further practice: he and the rest of his research team asked Dutch students to learn Dutch pseudowords (they were from an alien language, they were assured) via flashcards.[27] They were asked to do one of three things with the cards after each study round: put the happier-sounding words on a higher shelf and the negatively valenced ones on a lower shelf, place the positive word on the lower shelf, or simply place it on the neutral surface of the desktop. Pairing the happier word with the upward motion produced results. This "metaphor congruity"—putting positive words up and negative words down—led to a 4 percent bump in accuracy of identifying the words on a later test, enough to bump an A minus to an A in the standard American grading system.

But that original marble experiment's finding, compelling as it was, only brought up a deeper mystery: Why would upward physical movements be linked to happier affect in the first place? "Nobody knew where the mapping was coming from," Casasanto explains. Somewhere down the line metaphors got infused in the body. Or maybe it was the other way around. Two main camps had impressive arguments. The Lakoffians (followers of Lakoff!) contended that this happens because of correlations between bodily and emotional states: you stand up when you're feeling good and slump when feeling bad, and that principle of posture generalizes across experience, all the way into culture and metaphor. The psycholinguists, that deterministic camp that Casasanto flirted with earlier in his career, offered an alternate explanation: you don't need to discover this mapping by way of bodily experience; language gives it to you. You were born into a culture where you can't help but use expressions like "feeling up" or "feeling down." You have to use space to talk about both space and valence, or the emotional quality of things. Both arguments were compelling, but there wasn't a clear winner. He thought about it for years, and just couldn't find a way to detangle these spatial, emotional, linguistic admixtures of metaphor. "So," Casasanto told us in an interview, "I despaired."[28]

Then came a breakthrough. He hit upon the idea that in the English language and Anglophone culture, not only is *up* associated with good, but so is *right*. You want to have a right-hand man and avoid left-handed compliments, put the right foot forward and avoid having two left feet. That's a pattern found not only in English but in lots of languages—even dead ones like Latin. "The Latin words for *right* and *left, dexter* and *sinister,* form the roots of English words meaning 'skillful' and 'evil,'" he noted in a later paper, while the French (*droite*) and German (*Recht*) words for *right* are related to legal privileges, while the terms for *left* (*gauche* in French and *links* in German) relate to the words for distasteful or clumsy.[29] Indeed, *gauche* has been loaned to English as "lacking social experience or grace." Even the English word *awkward* stems from the Middle English *awke,* which means "turned the wrong way" or "left-handed." That idiomatic tic is reinforced by cultural mores, like swearing to tell the truth with your right hand or entering a mosque on your right foot. In many cultures, you're not supposed to point or eat with your left hand, since depending on where you are, that hand is reserved for some rather dirty jobs. So why would this be? Perhaps, because most people—that estimated 90 percent or so—are right-handed, and that imbalance has animated language and culture.

Relatedly, for a quarter century, psychologists had been finding strong links between fluency—how easy, accessible, or effortless something is—and evaluation, or how much you like or dislike a given thing. Fluency is a place where what's traditionally thought of as cognition and what's traditionally thought of as perception bleed into each other—it's the *feeling of* thinking and perceiving.[30] In broad terms, people like things that they can perceive or interact with more fluently, readily, easily. Remarkably, people will rate fluent statements as more true, more likable, more frequent, and more intelligent.

Putting these things together, it all clicked into place for Casasanto: fluency could explain the associations between right and left with good and bad. "With these lopsided bodies, we have no choice but to go through life interacting with things more fluently on our dominant side and uncomfortably on our nondominant side," Casasanto explains. "As we do, perhaps we come to associate dominant side with positive and nondominant with negative."[31] We feel fluent and good with our dominant hand, awkward and strange with the other hand.

So Casasanto and his colleagues followed up on the marbles study with more experiments to trace the links between handedness and goodness.[32] Some of the experimental designs were rather fanciful, like giving participants a sheet of paper with illustrations of two alien creatures—one on the left side, one on the right side. The researchers then asked the participants to assign adjectives to the aliens, like which looked more honest, attractive, and intelligent. Reliably, participants attached the positive attributes to the extraterrestrial on their dominant side. In another experiment, Casasanto had his participants put a bulky ski glove on their dominant hand, and then tasked them with carefully placing dominoes in a particular pattern. After just 12 minutes with the ski glove on their dominant hand, they put positive descriptors, like in that label-the-alien experiment, on the side of their nondominant hand. Just a couple minutes of fumbling with the ski glove was enough to get the nondominant hand to be associated with goodness.

"Unlike 'good is up,' which is true both in language and in the body, 'good is left' is completely absent from language and culture," he says.[33] "Lefties don't get to say that the correct answer is the *left* answer; they don't get to shake hands with their left hand, and swear to tell the truth and nothing but the truth with their left hand. They have to talk and act like righties. If these mental metaphors, these nonlinguistic mappings in our heads, were generated by language, then everybody should think 'right is good.' It's only if this mental metaphor is created through bodily experience that righties should think right is good and lefties should think left is good, in spite of everything language and culture is telling them. And that's the answer we get over and over." Casasanto and his colleagues have taken this finding out of the lab and into everyday life, revealing how our dominant hands, without our noticing, have a way of subtly guiding our behavior. He's found that presidential nominees—the righties George W. Bush and John Kerry, the lefties Barack Obama and John McCain—gesture with their dominant hands when praising something and their other hands when making a critique. And most astounding of all, Casasanto has found that English, Dutch, and Spanish speakers prefer words heavy with letters on the right side of the keyboard, and that, since 1990, Americans have been preferring baby names that begin with letters from the right side of the keyboard. From 2010 to 2018, some of the most popular first names were Noah, Liam, and Mia.

In the 21st century, we are not that far from the cave artists tracing their hands on walls dozens of millennia ago. Indeed, given the way that the typed speech of texting and emails have and continue to displace spoken speech, our hands mediate our interface with the world even more than they did in premodern eras. Steve Jobs, a pioneer of the user experience, once quipped that if he ever saw someone use a stylus with an iPhone, the project would have failed—the most intuitive user interface is the one that capitalizes on that instrument of instruments, the human hand. The hand shapes not only the way we see and plan actions but how we make judgments and meaning.

This is the borderland that the next part of this book will continue to explore: we have spent the last three chapters investigating how our bodies couple perceiving and doing. Next we turn our attention to knowing: how the decisions we make, the language with which we converse, and the emotions we feel are not all in our heads but are the fruits of how our bodies couple perceiving and knowing.

KNOWING

4.

THINKING

I N 1589, the Italian brothers Tommaso and Alessandro Francini moved from Florence to the village of Saint-Germain-en-Laye, the principal residence of King Henri IV, located just outside Paris. To honor the French king, the Francinis—engineers previously under the employ of Ferdinand I de Medici—would outdo themselves. For an expansion of the royal residence, the brothers designed a series of grottoes, partly enclosed galleries set on terraces leading down to the Seine. Water from the river would feed a fountain at the top of the terraces and then stream down to power a range of mythologically inspired scenes comprised of automatons: a cyclops in repose playing panpipes; a nymph playing an organ; Bacchus slugging back wine; Mercury blowing a trumpet; and Perseus descending from the ceiling and drawing his sword to slay a dragon that rose from a basin of water, thereby freeing Andromeda. When a passerby would step on a stone in the walkway, the weight caused a valve to open; this changed the water pressure in pipes within the mechanical figure, bringing about its lifelike movement. Long before artificial intelligence became a buzzword, automata were showing the European upper classes how machines could seem alive. In 1614, a peripatetic philosopher moved to town, a young René Descartes, who was inspired by what he saw in these royal gardens.

While he lived in Saint-Germain-en-Laye for only a year or so, these mechanical figures clearly had a major influence on the thinker. In his *Treatise on Man,* Descartes gives a long description of the functions of the human body, and he argues that everything from the digestion of food to the beating of the heart to the movements of the limbs to the actions of the brain can be understood mechanically. "I should like you to consider that these functions follow from the mere arrangement of the machine's organs every bit as naturally as the movements of a clock or other automation follow from the arrangement of its counter-weights and wheels," he wrote. "The nerves of the [human] machine that I'm describing can indeed be compared to the pipes in the mechanical parts of these fountains." This led Descartes to what, to him, distinguished man from animal: his rationality. A human is "a thing that doubts, understands, affirms, denies, is willing, is unwilling, and also imagines and has sense perceptions," he added in his *Meditations on First Philosophy.*[1] "When a rational soul is present in this machine it will have its principal seat in the brain and reside there like the fountaineer, who must be stationed at the tanks to which the fountain's pipes return if he wants to initiate, impede, or in some way alter their movements," he wrote.[2]

To Descartes, the ghost in the machine of the human body was a transcendent intellect, the soul. This was the latest in a long tradition in separating man from animal. Going back to Aristotle, it was thought that our soul resided in our intellect. This soul was eternal, divine, and separated *us* (humans) from *them* (animals). Indeed, it was doctrine for medieval Christianity that identifying with the body—and its passions—was the very definition of sin. This was the accepted cultural wisdom of Descartes's day.

Descartes gave us what is now called Cartesian dualism, a philosophical stance that asserts that people are comprised of both a physical body (including the brain) and the immaterial mind. You can spot this separation of body and mind in everyday language use. We talk about "having bodies" but rarely "being bodies." In this conception, the body is a vehicle that carries the mind about.

That we feel comfortable separating mind from body may reflect inherent biases of human thought. All of us walk around with a system for dealing with physical objects and another system for social entities.[3] Paul Bloom, a Yale psychologist and philosopher, has noted that mind/body dualism comes naturally to children. He observes that young children "will tell you that you need your brain for certain actions, such as solving math problems, but not for others, such as loving your brother, or pretending to be a kangaroo."[4] For children and adults, it seems quite natural to suppose that there is a realm of things that the body does, like walking and doing mental arithmetic, and another, more ethereal, realm that is the provenance of mind. This latter realm is responsible for hard-to-explain human propensities, such as love, creativity, and astonishment.

The brain is, of course, a physical device, if we may briefly call it that; and throughout history an astounding number of metaphors have been employed to describe its operation, each taking inspiration from currently prized technologies of the day. Inspired by the Francini brothers' automatons, Descartes described the brain as a hydraulically powered machine. Others have compared the brain to a wax block (Plato), a blank slate (John Locke), a mill (Gottfried Wilhelm Leibniz), a hydraulic system (Sigmund Freud), a telegraph (Hermann von Helmholtz), and a telephone switchboard

(Charles Sherrington, who won a Nobel Prize for his work on neurons). As of this writing, in psychology, cognitive science, and the popular beliefs surrounding them, the prevailing metaphor is the computational theory of mind. The brain is metaphorically compared to a computer's hardware, the equivalent of a CPU and hard drive, and the mind is its software, an operating system furnished with many applications. This reasoning by analogy—that your brain is a computer and the mind is its operating system—is today's accepted wisdom for how the body and mind relate to each other.

It's one thing to say mental processes can be modeled with computer programs, and thereby provide a means to predict behavior. This endeavor is good science and akin to weather forecasting, which also uses computer programs to model the weather. However, it is quite another thing to say that the mind or the weather is a computer program. In the case of the weather, this would be a silly assertion. Yet, the computational theory of mind does indeed make the strong claim that the brain is a computer and the mind is its software. Computer models are an extremely useful means to understand, simulate, and predict cognitive performance. But this does not mean that the brain is a computer. Computer models are also a useful means to understand, simulate, and predict the weather. But the weather is surely not a computer program, and simulated rain won't get you wet. A brain is a bodily organ that evolved to support the ways of life of a particular organism. Computers are man-made, lifeless artifacts.

GUT FEELINGS

IN THE SPRING OF 2012, a team of researchers at the universities of Cambridge and Sussex in the United Kingdom sought out a rather exceptional cohort of participants for an experiment—hedge-fund traders in the City of London, one of the world's financial capitals.[5] These eighteen financiers, all men, practiced a vertigo-inducing form of the craft, making trades in seconds

THINKING

or minutes, or, in more long-term deals, over a period of hours. It's up to the trader to infer patterns on the fly; make sense of huge amounts of rapidly shifting data; and, as the researchers note, make major decisions in a matter of seconds. What's more, their compensation was structured so as to maximize individual responsibility—there was no end-of-year, firm-wide bonus; instead, traders were paid in proportion to their trading success. And while a given trader could make a ton of money based on his individual performance—approaching £10 million (approximately $13 million)—in a single year, he could also be bounced out of a job quickly. Compounding these factors, the researchers came to them at the close of what's come to be known as the European sovereign debt crisis, where, in combination with the wider global financial crisis, countries on the periphery of the Eurozone economy were tottering toward collapse, with a country like Greece going into extreme national debt. All this is to say that the jobs of these traders were extremely demanding; extremely lucrative; and happened to be, at the time of study, in an extremely volatile state. So the researchers asked them to listen to their hearts.

First, each trader was asked, while sitting silently, without touching his chest or his wrist or anywhere else where the pulse is palpable, to count how many heartbeats he had over short lengths of time—25, 30, 35, 40, 45, and 50 seconds—done in a random order. All the while, a monitor was recording his heart rate, thereby allowing the researchers to compare the trader's perception of his heart rate to the actual number of beats in a given period. After each of the trials, the trader then rated how confident he felt about his estimates, from their being total guesses to having complete confidence in their accuracy. Following that, the trader then carried out an auditory version of the test, listening to tones that were either synced with his heartbeat or came at a delay. After each round, he was asked if the tones were synchronized or not, repeating the process for a total of 15 rounds. The researchers also gathered the traders' ages and how long they'd been trading, as well as

81

a standard metric of business success—their individual profit and loss statements, which score how much money they made or lost in a given year. The researchers did the same tests on an age-matched group of nontrader men drawn from the University of Sussex.

The traders, the researchers found, were far better at reading their own heartbeats than the laypeople from the university.* What's more, among the traders, being more accurate was linked to greater profitability. Accuracy in heartbeat detection was also predictive of how long a trader had survived in the financial markets. Perhaps most intriguingly, as a group the experienced traders were much more accurate at detecting their heartbeats than were the beginners. And while the accuracy of the heartbeat sensing was predictive of performance, a given participant's confidence about how they did on the accuracy test didn't matter at all. "Our results suggest that economics and the behavioral assumptions upon which it rests, will benefit from a greater involvement with human biology," the researchers wrote in the close of the paper. "Today there is a debate between, on the one hand, classical economists who argue that economics has no use for the findings of psychology and neuroscience, and on the other, behavioral economists who do draw lessons from these experimental sciences. What has been missing from this debate is evidence concerning the role of somatic signals, in other words, the body, in guiding our decision-making and behaviour and, crucially, our risk taking."[6]

The researchers were interested in a perceptual process called *interoception*, or how people sense their own internal states. It's an interoceptive signal at work when you realize that your stomach is full and you should stop eating or that your bladder is full and you need to empty it. Interoception tends to operate in the background of your awareness unless something

---

* For each participant, a heartbeat accuracy score (HS) was computed. For each trial: HS = $1-[|nbeats_{real}-nbeats_{reported}|] / [(nbeats_{real}-nbeats_{reported})/2]$. This score was averaged across trials and participants.

needs to be acted upon—it's unlikely you're noticing digestion until indigestion hits. (There are exceptions. As a runner, Drake has, over time, gained a finer-tuned sense of the progress of his digestion, so that meals and runs can be timed for maximum fuel and minimum load, with trotting out for a jog around three hours after eating). More generally, we don't often notice that our internal body is being continuously sensed unless something is amiss, like becoming nauseous or exhausted. Interoception is sensing turned inward; at all times, receptors from your tissues and organs are sending signals up to the brain, including from the heart.

Not unlike how visual experience isn't the sole domain of the eye, thinking is not all in your head. To wit, every time the heart beats, it activates pressure-sensitive neural receptors in the heart, which detect and send signals about heart functioning to the brain. At all times, the brain is in dynamic communication with the heart, other organs, and the rest of the body. "Every moment, our brains are representing the activity of different organs," says study coauthor Sarah Garfinkel, a professor of psychiatry at Sussex.[7] "This can serve to influence how we think and feel about the world. For me, this marks a shift in how we approach neuroscience traditionally, to take a more embodied perspective, to see the brain as embedded within the body."

The traders in Garfinkel's study are sensing their internal milieu as it shifts in response to the perceived movement of the market. These "gut feelings" are taken as a good/bad evaluations of market trends. In earlier chapters, we saw the same pragmatic principle at work with perceiving the slope of hills: if you're well resourced, the hill looks easier to ascend. A hill looks inviting or foreboding, and so does a stock purchase. Perception guides action.

Garfinkel has investigated interoception in a range of contexts. She's found that the bigger the gap between someone's confidence in reading their heartbeat and their actual performance on heartbeat self-measurement tests, the more anxiety they feel in everyday life. The takeaway: being disconnected from your body can lead not only to worse decisions but also to greater anxiety. That also suggests that interoception may be a possible access point for addressing anxiety and related issues: training people to attend to their body's internal signals has been shown to increase interoceptive accuracy,[8] including the accuracy of self-assessments of blood-glucose levels in patients with type II diabetes.[9] The implication: the more attuned you are to your bodily states, the better the decisions you'll make, whether it's purchasing stocks or taking care of your health.

THOUGHT AND BIOENERGETICS

WE ARE ALIVE. The reality and consequences of this are astonishingly absent in almost all of psychology, both pop and academic. As living beings, we are driven to pursue two biological imperatives: survival and reproduction. Although we are rarely aware of it, these imperatives underlie almost everything that we do.

Survival is largely driven by bioenergetic imperatives. Eat more calories than you expend. In addition, there are imperatives to seek shelter from inclement weather and avoid getting eaten. Satisfy these imperatives and you'll survive.

For living beings, doing requires energy. In our previous three chapters—doing in the form of developing, walking, and grasping—we focused on how we see the bioenergetic costs of acting in our surroundings. Hills and distances are perceptually scaled by the bioenergetic costs of locomotion. Energy is the price we must pay in the marketplace of action selection. It is a currency of life. Everything we do entails a cost or benefit to our energy economy, including thinking.

Thinking requires energy and is subject to bioenergetic considerations. Brains are both consumers, hoarding 20 percent of human resting-state metabolism, and guardians of energy resources. As guardian, the brain must seek and manage energy in both a feed-forward and a feed-back manner.

Feed-forward systems anticipate the future. We typically don't go to the grocery store because we're hungry. We go to the grocery store to avoid becoming hungry. Feedback systems respond to the present. Being hungry is an unpleasant feeling that goads us into finding food fast—raid the refrigerator or pantry now! The computational theory of mind, which assumes that the brain is a computer, has no truck with life's imperatives. Computers are not living beings; they are not subject to life's imperatives. Computers can certainly simulate biological processes—much as they can simulate the weather—but computers are no more alive than they are thunderstorms. Life's imperatives are realized in an ecological marketplace in which the currency is energy.

This helps explain why something like blood sugar levels can affect the way that we evaluate things. Researchers Xiao-Tian Wang and Robert D. Dvorak, then at the University of South Dakota, investigated this through the phenomenon of future discounting, where people reliably favor immediate rewards over larger rewards coming in the future.[10] For example, most people would rather accept $100 now rather than wait a week and receive $105. In their experiment, Wang and Dvorak manipulated blood glucose levels with the help of the soft drink Sprite. In the study, participants were asked whether they would prefer to take less money tomorrow (between $90 and $570) or more money later (4 to 939 days). Participants answered such questions twice, before and after drinking a glass of Sprite that, depending on experimental group assignment, contained either sugar or a noncaloric sweetener. The result: the participants who drank the sugary drink were more willing to delay gratification and take more money later. The *Sprite*—or the sugar within it—defrayed future discounting. Similarly, studies[11] indicate that people will be more helpful after they've enjoyed a tall, sugary glass of lemonade.

Another study looked at thought in the wild by investigating the decisions of parole judges in Israel. These judges must decide on whether to grant parole to prisoners who are petitioning for release from prison. By far, the most common decision across judges is no. There are, however, exceptions to the likelihood of this negative outcome. All petitions are heard during a long morning session that is broken up by two breaks in which the judges retire and have a snack. First thing in the morning, and after each of these breaks, the judges are highly likely to grant parole; but with time and fatigue, their decisions become increasingly negative. When their fuel tank is on empty, judges fall back on the easy default of "just say no." After a snack, they are more likely to consider the case more deeply and arrive at a positive ruling. Like any other physiological process, judicial decisions require fuel.[12]

All this has important policy implications, especially in education. A recent study of California public schools found that students on average scored about four percentage points higher on end-of-year academic tests when their schools had contracted with a healthier school lunch service.[13] Indeed, providing up to three meals a day looks to be a key ingredient in high schools that graduate more low-income students.[14] Similarly, children who eat breakfast at school have better attendance records and fewer behavioral incidents.[15] If being momentarily undernourished causes judges to make less thoughtful "just say no" rulings, then imagine what chronic food insecurity does to the cognition of eight-year-olds. Whoever you are, thought is influenced by current physiological energy resources. The way you think is endlessly tied to how you physically feel—and how you read those feelings.

FLUENCY

IN 1991, NORBERT SCHWARZ, a psychologist now at the University of Southern California, and his collaborators asked dozens of German college students to recall times that they felt assertive or unassertive. Some had to think of six examples, others had to think of a full dozen.[16] They were also asked to rate how easy it was to come up with those examples, as well as rate how assertive or unassertive they felt themselves to be. Here's the surprising result: those participants who were required to recall just six examples rated themselves as *more* assertive than those prompted to come up with a dozen examples. In essence, the more examples of assertiveness that someone had to recall, the less assertive they thought themselves to be. Why would this be? It is because it is easier to think of a few examples of when you were assertive than to think of many examples, and you judge how assertive you are based upon how easy it is to think up examples. A few examples, easy-peasy. Many examples, much harder.

It's tempting to assume that cognition is, underneath it all, nothing but computation. So says the computational theory of mind. But one of the many things that the computational model glosses over is the *feeling of thinking,* or how we perceive our own thoughts, helps determine the thoughts we end up having. For instance, in the last chapter, we explored the many ways that the ease of physical movements shapes judgment. Easy actions feel good, so we say that they are good. As we discovered last chapter, people like things more when they are on their dominant side, hence "the right hand of God" and "left-handed compliments." Schwarz's research shows that the same goes for thinking. Degrees of ease and effort frame how accurate, true, and confident people feel about their own thoughts, as well as statements from others. Like physical activities—moving a refrigerator, going for a run—thinking is accompanied by feelings of effort. Happening largely out of the scope of attention, these feelings guide the ways we evaluate things. We are not automata or computers, we are living beings. And the way it feels to think guides our thinking. Moment by moment, we are perceiving our own thoughts; otherwise we'd never have access to them.

Thinking is an embodied activity, just like playing a sport. Sometimes it goes fluidly and sometimes it is effortful. In sports, being "in the zone" means your body is doing what you want it to do with minimal assistance or interference from the conscious, judgmental part of yourself. Not being in the flow means that things are not going well, a state that is typically accompanied by unproductive thoughts of self-coaching and self-loathing. Whether moving or thinking, the feeling of effortless fluency tells you that you are good at what you are doing, or at least you think you are.

Just as physical actions range from being easy to hard, so do mental ones. Fluency can be perceived in many contexts, like when noticing how easy it is to read well-written prose, how nice it is to listen to an engaging melody, or how easy it is to remember a phrase. Without our being consciously aware of it, the fluency of an experience guides our value judgments. Fluent writers strike us as being intelligent and fluent statements sound true. And as one clever yet grim experiment found, when people who speak English as a second language with a heavy accent are asked to repeat statements of trivia, they're less likely to be judged as believable as people speaking in their native tongue. If someone tells you, "A giraffe can go without water longer than a camel can" (which is true!), you'll be more likely to believe them if they speak without an accent—even if they're just playing the messenger and saying a thing provided to them,

Experiments have shown how easy it is to manipulate this fluency-truth effect. In one early study, researchers asked participants about statements like "Lima is in Peru" while manipulating ease of reading by tinkering with the contrast and color of the printed text against its background. The statements were more likely to be deemed true when they were easy to read, like red text against a white background than if they were difficult to read, like yellow or light blue against that same white background. Rhyming has been found to be a marvelous means to promote fluency. In these studies, participants rated aphorisms that rhyme as being more accurate than those that don't: "Woes unite foes" versus "Woes unite enemies" and "What sobriety conceals, alcohol reveals" versus "What sobriety conceals, alcohol unmasks."[17] But, it's important to note, this effect was reduced when participants were asked to separate the truth of a statement from its poetic value.

The power of rhyming to facilitate memory, and thereby imbue a sense of truth, is a cultural universal. The rhyming epic poem is a centerpiece of oral traditions since, scholars contend, it's much easier to recall verse tied together by rhyme and rhythm than dry, bloodless prose. Friedrich Nietzsche thought that rhyming was a conduit to memory and a device with which people reached toward what they felt was divine. "The rhythmic prayer seemed to get closer to the ears of the gods," he wrote in *The Gay Science*.[18] More recently, recall the statement about the bloody glove at the trial of O. J. Simpson: "If it doesn't fit, you must acquit." But here's a fascinating twist: *tell* people that rhyming is having an effect on their judgment, and the spell will lose its power.

The role of fluency in human judgments is an extension of what the legendary experimental psychology team of Daniel Kahneman and Amos Tversky would call the availability heuristic back in the 1970s.[19] The *availability heuristic* relates the likelihood of something occurring with the ease with which it can be brought to mind. Try this example from one of their studies: Are there more English-language words with the letter *K* appearing as the first or as the third letter in the word? That is, are there more words like *kangaroo* or *acknowledge*? In general, people think that there are more words with *K* in the first position, *kangaroo,* than in the third, *acknowledge*. In fact, the reverse is true. Why are people so wrong about this? Because, naturally, it's much easier to think of words that begin with a given letter than those with a letter in the middle of the world. Try this: think of five words beginning with *K* and then think of five words for which *K* is the third letter. Which was easier? Words beginning with the letter, of course. The ease of recall serves as a substitute for frequency of occurrence. This has massive political and social implications. As Kahneman has noted, over a wide range of topics, our sense of what's common or rare in the world is shaped by media, including social media. Take, for instance, the inflammation of fears of foreign terrorists that animates so much political messaging, particularly from the American right.

Since the September 11, 2001, attacks, foreign-born terrorists have killed, on average, about 1 American per year. Add US born, and 6 Americans die per year to Islamic terrorism.[20] Relatedly, reports indicate that an American is 29 times more likely to die from an asteroid striking them than at the hands of a refugee terrorist, and 129,000 times more likely to die in any firearms assault, and 6.9 million times more likely to die of heart disease or cancer.[21] Media theorists have a name for this sort of thing: mean world syndrome, where the violent content of so much mass media leads people to think that the world is more violent than it actually is. To wit, the American crime rate has fallen precipitously since its 1991 peak, yet nationally representative surveys indicate that Americans *think* that the crime rate has gone up every year since the early 2000s. Why? Well, children will see an estimated 8,000 murders on television before they turn 12 years old.[22] This also speaks to how important positive media representation can be for shifting attitudes. In one recent study, 193 white Americans were asked to watch one of two shows: *Friends,* the NBC megahit that has since gained notoriety for its staggering lack of diversity, or *Little Mosque on the Prairie,* a Canadian sitcom about a Muslim family living in a small town in Saskatchewan. After watching, the *Mosque* viewers expressed more positive attitudes toward Arabs compared to those who had watched *Friends.* Importantly, identifying more with the out-group members was linked to greater reductions in prejudice.[23]

In many contexts, fluency can be a false friend. For children and adults, fluency can actually *impede* learning. For example, when studying for an exam, many students will rely on highlighting and reading over notes, since it is easy to do and makes you *feel* like you've got a good grasp on things. But long-term knowledge is actually built through active recall, which is much harder and feels dysfluent: quizzing yourself before and while studying feels terrible, but loads of studies indicate that it leads to better learning. It's a similar case for diversity and inclusion in organizational settings: homogenous groups—all 20-something white males working in a small startup, for instance—*feel* fluent to their members, which leads to unreflective (and rash) decision-making.

In experiments, the *dys*fluency engendered by diverse teams leads to more thoughtful decisions. Sitting at a table with people who look, sound, and think differently from you is naturally less comfortable than sharing a space with people who have similar backgrounds to yours. That discomfort leads to more and stronger reflection. Studies from Jamie Pennebaker's Language Lab at the University of Texas suggest that fluency is key to winning people over to your side of an argument. Graduate student Ryan Boyd, now as assistant professor at Lancaster University in the UK, analyzed the results of Intelligence Squared, a debate series where teams argue before a live audience that then votes for who they think won. Boyd found that the more complex the language debaters used, the less likely they were to sway people, and the more concrete the language, the more convincing the argument. "Things that require no effort and are just intuitively understandable and relatable—people inherently like that," he explains.[24]

Thinking does not exist apart from feeling. Our perceptual world is always filled with the feelings of effort attendant to action and thought. We take from these feelings cues about what's right, what's wrong, how likely something is to occur, and ultimately what the structure of society is like. Which is to say, thinking is a bodily experience. And thought itself, even in its most abstract forms, is never divorced from the feelings inherent in being alive.

PUTTING FLUENCY—AND GROUP DECISION-MAKING—TO TRIAL

THINKING, AS YOU CAN tell by now, isn't simply something that happens in your head. Or your body. The long tradition of social psychology has shown how the groups we find ourselves in affect the decisions that we make. This was brilliantly investigated by Samuel Sommers, a psychologist at Tufts, who investigated the influences of group composition in a real-world contexts for what would become a landmark 2006 study.[25] He worked with a county courthouse in Washtenaw County, Michigan, near Ann Arbor and the University of Michigan, where he earned his PhD. Working with local judges and jury-pool administrators, Sommers sourced 200 participants to form what ended up being 29 mock juries of 6 people each. Half of those juries was ethnically homogenous—as in, all white people—while the other half was ethnically diverse—2 black people, 4 white. About 60 percent of the participants were women.

These jury groups were then seated at rectangular courthouse tables and assigned juror numbers, sitting together and in plain sight of one another. They then went through voir dire, where potential jurors are weeded out due to bias. The questioning in one version was race neutral, while the other addressed attitudes about race. Then the group watched a 30-minute Court TV[26] television summary of a trial where a black defendant was charged with sexual assault—it had highlights of the opening and closing arguments from the defense and the prosecution and witness testimony. In the video, the prosecution's argument hinged on forensics: the semen and hair found on the scene were consistent, but not definitive, matches for the defendant. The defense countered that the forensic evidence was inconclusive and that there were no eyewitnesses.

The jury was then asked to deliberate on the trial that they had just seen on TV and their discussions were videotaped. A foreperson was selected, who would speak on the jury's behalf. If a consensus was achieved, then the foreperson would alert the experimenter and report the jury's decision; otherwise deliberations would be capped after an hour.

The outcomes were remarkably similar: only 1 of the 29 mock juries reached a unanimous guilty verdict, and that was an all-white group. There was a roughly even split between the diverse and nondiverse groups in terms of acquittals; and the diverse groups were a little more likely to result in a hung jury, but only for those who had gone through the race-relevant voir dire questioning beforehand. What's striking, however, is *how* they came to their decisions: the diverse groups took a full 12 minutes longer to reach a decision, at 50 minutes on average; they considered more case facts, made fewer inaccurate claims, and were more likely to mention racism. But these extended deliberations weren't simply because the people of color brought up more perspectives—rather, the white people in these diverse groups were more circumspect. White people in diverse groups were more open to talking about race, and raised more case facts in general. Sommers highlighted a couple exchanges between jurors that illustrated the different habits of deliberation that emerged. In the all-white groups, racism, when brought up, was dismissed as an irrelevant topic. *(Juror 3: "But I'm telling you, you know? I'm sorry . . . [the victims] can't tell one Black person from another?" Juror 6: "I don't buy that.")*[27] But in diverse groups, white members directly appealed to black members to validate their claims. *(Juror 6: "We're going through that right now with the terrorism. They're, you know, looking at people of Arab descent pretty cautiously. And, and I think they [law enforcement] do a lot of that, especially with Black males. I don't know how often you guys (gestures to Jurors 2 and 3) run into this, but . . ." Juror 3: "I, I grew up with it, racial profiling.")*[28] That's the big takeaway: people act more thoughtfully when they're not surrounded by people who look like them. "One of the ways in which White participants' performance varied by group composition was that they made fewer inaccurate statements when in diverse versus all-White groups, despite the fact that they actually contributed more information when deliberating in a diverse setting," Sommers wrote. "This result suggests that White jurors processed the trial information more systematically when they expected to deliberate with a heterogeneous group." Hence, putting yourself into a diverse group might make you feel uncomfortable, but it's going to lead to more thoughtful decisions—in the case of this particular study, it induces you to seek and consider more evidence.

FLUENCY IN THE WORLD: BULLSHIT

THAT SO MUCH OF how and what we think is shaped by embodied, phenomenological forces also makes us profoundly susceptible to bullshit. The corrosive ubiquity of bullshit was first called out by the Princeton philosopher Harry Frankfurt, in a philosophical essay that became a surprise number one *New York Times* bestseller.[29] His *On Bullshit* begins without pulling punches: "One of the most salient features of our culture is that there is so much bullshit. Everyone knows this." The difference between the mere liar and the bullshitter is one of regard toward truth. The bullshitter's statements are "unconnected to a concern with truth," he writes; they are not "concerned with the truth-value" of what they are saying. The liar, one may charitably say, at least respects the truth enough to obscure it. The bullshitter pays it no mind.

Frankfurt, who published his text in 2005, had a bead on a trend that would only become more evident and powerful in the intervening years. The presence of bullshit, which he also more genteelly refers to as *humbug,* can be realized in any number of ways. For example in *The Colbert Report,* the comedian Stephen Colbert, coined the term *truthiness,* which refers to something that feels like truth, but doesn't have to be yoked to the factual.

Truthiness. Just a few months later, it would be chosen by the American Dialect Society as the 2005 word of the year, beating out *podcast, lifehack, sudoku,* and *cyber Monday.* Its introduction coincided with the beginning of a divorce between empirical evidence and what was taken to be a fact, reinforced by a decline of journalistic institutions and a growing prevalence of social media. (In 2005, just 7 percent of US adults used social media; it was 65 percent by 2015.[30]) Who could have predicted that Facebook—launched in 2004—would become a platform for pushing bullshit, humbug, and truthiness at such a scale that it could help swing a US presidential election a dozen years later?[31] Our penchant for fluency makes us susceptible to bullshit—if it feels right, it is right—and when that vulnerability is scaled up to the level of media, you get truthiness fake news.

The production of bullshit seems to come down to ease and lack of accountability. In one experiment, Wake Forest psychologist John Petrocelli recruited more than 500 online participants to answer questions about a fictional local politician named Jim, who had just pulled out of the race for city council.[32] The participants were asked to come up with five possible reasons for why Jim decided to throw in the towel. Petrocelli tinkered with the conditions for bullshitting in several ways: some participants were given actual background knowledge about Jim in the form of autobiographical facts; some were told they didn't have to supply answers if they didn't want to; others were told that their answers would be reviewed by a panel of judges who knew Jim quite well. Petrocelli later asked participants to report how concerned they were with evidence in making their claims—a self-diagnosed bullshit score. All these factors influenced how much bullshitting occurred: people who didn't get any background information about Jim were heavy bullshitters, and so were those who felt they had a social obligation to come up with explanations. That is because, Petrocelli reasoned, people bullshit when they feel obligated to communicate about things they actually know little about; and so when they think they can slide by on a "social pass of acceptability" for doing so, they'll opt to bullshit. Like so much of human behavior, bullshit is highly social: if you think you can get away with it, you're more likely to bullshit. But when the social cues around you signal that it'll be hard to get a free pass—as was the case for those who believed that their answers would be judged by experts—then you'll be much less likely to make things up.

A follow-up asked participants to explain their rationale behind opinions on nuclear weapons, affirmative action, or other hot-button issues, with some being told that a sociology professor would be evaluating their claims. In the experiment, those who didn't hear about the professor or thought that he agreed with them were most likely to bullshit. (Some of the responses in the no-accountability condition included: for affirmative action, "It partly helps to solve the problem of unemployment," and regarding capital punishment, "Imprisonment for life has the same effectiveness as the death penalty.") Hence what Petrocelli coined as "the ease of passing bullshit hypothesis"— fluency, once again, enables the generation of bullshit. If it feels easy to bullshit, you're more likely to bullshit.

Fluency also drives whether or not we fall for bullshit as perceivers. Research psychology has identified this as the "illusory truth" effect: the more one is exposed to a false statement, the more true it appears to be. In lab settings, participants were exposed to a range of statements that are true ("A prune is a dried plum") or not ("A date is a dried plum"). Then, after a break of minutes or sometimes days, they were asked to evaluate another batch of statements, including some they'd seen before. And reliably the false statements they had been exposed to earlier were rated as more true.[33] And a related literature on "knowledge neglect" suggests that it doesn't even matter if you actually know the subject matter being assessed: false statements read twice are frequently given higher truth ratings than novel statements, even when both sorts of statements conflict with prior knowledge. There is a hopeful takeaway, however: when participants are asked to explain *why* a statement is true or false, the illusory truth effect evaporates. "Inferring truth from fluency often proves to be an accurate and cognitively inexpensive strategy, making it reasonable that people sometimes apply this heuristic without searching for knowledge," writes Lisa Fazio and her coauthors in the study's conclusion.[34] Fluency makes us vulnerable to mistaking ease for truth—it's a function of how we perceive our own thoughts. But with a little inquiry, the spell can be broken.

As we've seen, thoughts are embodied. They appear in our perceptual world—our *Umwelt*—colored with feelings emanating from within our body. Thoughts are accompanied by gut feelings, feeling of effort, affect, and emotions. Reason and emotion are not distinct, separable faculties of the mind. Rather, they are knitted together in surprising ways. In the case of emotions, affective feelings actively cajole us to "stop" or "go"—to proceed with our endeavors or to withdraw. Our feelings are enmeshed in our reasoning about what stocks to purchase, who to fear, and which party to vote for.

5.

FEELING

THE FOLLOWING ITALICIZED SENTENCES are meant to be distasteful. They are items from the disgust sensitivity scale (DSS) that, as its name suggests, is designed to assess individual differences in people's experience of disgust when confronted with repulsive things.[1] Some items ask you to state how strongly you agree with the following statements: *I might be willing to try eating monkey meat, under some circumstances. If I see someone vomit, it makes me sick to my stomach. Even if I was hungry, I would not drink a bowl of my favorite soup if it had been stirred by a used but thoroughly washed fly swatter.* Other items ask you to rate how disgusted you would feel when confronting the following experiences: *You see maggots on a piece of meat in an outdoor garbage pail. While you are walking through a tunnel under a railroad track, you smell urine. A friend offers you a piece of chocolate shaped like dog-doo. You see a bowel movement left unflushed in a public toilet.* As silly and gross as this assessment scale may seem, since the mid-2000s, numerous studies have employed it, mostly in the United States but also in other countries. These studies have found that how easily you become disgusted—as measured by the DSS—is a reliable indicator of your politics.

People who are easily disgusted tend to be politically conservative, a finding that came as a surprise to the research team that discovered it. Yoel Inbar, then at Cornell University and now at the University of Toronto, was partnering with his mentor David Pizarro, a social psychologist interested in morality. The researchers had been looking into moral judgment and disgust, administering the DSS along with several other assessment scales, including a measure of political orientation. The link between disgust and conservatism stood out to Inbar and reminded him of how repulsive descriptors were—and largely, as of this writing, still are—often used to disparage gay people.

The initial experiment linking disgust and conservatism became a pilot study for what would become an influential paper by Inbar, Pizarro, and the philosophically minded Yale University psychologist Paul Bloom.[2] They recruited 181 participants from US swing states, finding again the link between disgust sensitivity and political conservatism. In subsequent research, Inbar and his collaborators found that disgust sensitivity was also associated with negative feelings about people who supported abortion and gay rights and also those who opposed gun rights.[3] In another study, Inbar and his colleagues found that disgust sensitivity was linked to voting for John McCain in the 2008 presidential election, and that across a whopping 121 countries around the globe, *contamination disgust,* which is concerned with germs and pathogens and other icky interpersonal things, predicted conservatism. Another found that students who had been surreptitiously exposed to a foul smell—by way of a novelty stink spray deployed in a laboratory trash can—felt less warmly toward gay men than a control group who did the same evaluation in a stench-free environment.[4] Other research teams extended the findings. Neuroscientists at Virginia Tech found that disgusting images, like of a used toilet or a mutilated corpse, prompted such marked neural responses in different participants that with just one image you could tell if they were conservative or liberal.[5]

Why should someone who is easily disgusted be more likely to be politically and socially conservative than someone who is not? Consider first that a conservative person is someone who is cautious about change and innovation. Conservatives value traditions that have stood the test of time. If it is not broken, then don't fix it. A liberal is less enamored of tradition and more apt to take risks. Food preferences exemplify the conservative/liberal stances. Some people like to stick with familiar foods, whereas others like to explore and taste new things. Each approach has its costs and benefits. By only eating familiar foods, the culinary conservatives maintain a safe diet—what they eat is highly unlikely to disappoint or poison them. The cost of this cautious stance is that they may miss out on foods that they would really like if they'd tried them. More curious eaters, on the other hand, are more likely to take risks, thereby getting to taste new foods but also exposing themselves to the dangers of foul flavors, indigestion, and food poisoning. Either stance can be preferable, depending upon the environment. In a world where everything is delicious and safe to eat, the liberal eater wins in the economy of calorie acquisition. Conversely, the conservative wins when tempting unfamiliar foods have a chance of being poisonous.

The objects of disgust are things that could make us sick: food, bodily wastes, and such. In the case of foods, if we do eat something poisonous and survive, then afterward the slightest smell of that food will make us nauseous and fill us with disgust. Such learned food aversions are nature's way of protecting us from becoming poisoned again. Disgust says *No!* in the most emphatic way possible. There is no way that you will eat something that smells deeply repulsive. The feeling of disgust is an evolved policing system meant to protect us from harm.

Compared with our hunter-gather ancestors, people today are rarely poisoned and instead are taught the targets of revulsion by emphatic parental and societal warnings. A really disgusting object is anything that will cause a parent to shriek at their child, "Don't touch that!" or especially, "Don't put that in your mouth!" The result of such parental and societal policing is the development of our own internal policing system: emotions.

Emotions are the police officers of the mind. They regulate our behavior by telling us, "Go! This is good," or "Stop! This is bad." This policing function has been co-opted by our value systems, such that certain immoral behaviors—for example, incest—may evoke feelings of disgust as vivid and strong as those evoked by the foulest of smells. Our sense of what is morally abhorrent is grounded in embodied feelings of disgust, feelings that originally evolved to protect us from things likely to make us ill. (Hence, *You make me sick!*) And, as we will see, we possess a palette of affective feelings— disgust, surprise, fear, happiness, etc.—with which to color our perceptual world. Operating in the background, our emotions are assessing the situations we're in, coaxing and cajoling us toward and away from people, places, and things. Emotions shape our personal worlds—and police who is welcome within them.

While controversial, this is not an entirely original insight. The Scottish philosopher David Hume, writing in the late 18th century, was skeptical of the "rationalism" that was popular among the philosophers of his day, which touted that you could arrive at truth by thinking alone, so long as it was sufficiently clear of emotion and other distractions. "Since vice and virtue are not discoverable merely by reason," he wrote in *A Treatise of Human Nature,* then it must be by some feeling or perception that we make sense of right and wrong. "Our decisions concerning moral rectitude and depravity are evidently perceptions," he observed. "Morality, therefore, is more properly felt than judged of."[6]

PERCEIVING EMOTION

IN 1999, DENNY WAS asked to participate in a congressional exposition of university research funded by the Department of Defense (DoD). Denny's DoD-funded research used virtual reality (VR), which was a new, expensive, and uncommon technology at that time. After considerable preparation building virtual worlds and portable scaffolding to hold the VR tracking antenna, Denny's lab—staff and graduate students—loaded up a rental van with the necessary equipment, drove up to Washington, DC, and set up a VR installation in a congressional office building. The virtual world that they created exploited one of VR's most engaging environments, where the user is placed at the edge of a very high cliff. This abyss from hell scenario was created for its entertainment value and had absolutely nothing to do with the research that Denny was doing with DoD funding. The DoD research was outlined on a poster adjacent to the VR installation, but no one looked at it. The VR installation stole the show.

After putting on the VR headsets, members of Congress and their staff were first placed into a virtual hall of Virginia presidents: Washington, Jefferson, Madison—eight in all. The congresspersons were then asked to approach and step onto an escalator in this virtual world, which ferried them up a couple flights of stairs. Then, when they got to the top of the escalator, they discovered the abyss from hell, a construction hole in the floor, cordoned off with yellow police tape. When the congresspersons looked over the edge, they saw many floors were missing. Were it a real environment, then falling into the hole would kill you. At the sight of the abyss, the members of Congress froze, exclaimed in distress, trembled, and often started giggling. The only escape to safety was to walk along a narrow ledge next to the abyss, which led to a down escalator. Denny guided them to the escalator, where they were conveyed down, removed their headsets, and shouted about how awesome was the experience they'd just had. What is most interesting about this story, for our present purposes, is that the members of Congress knew that they were safely standing on the floor the whole time and yet they still could not help but experience intense feelings of fear and anxiety about the possibility of falling into the virtual abyss. You might think that their knowledge of where they really were would have prevented them from being afraid, but it did not. How come?

To answer this question we need to introduce a little functional neuroanatomy. The brain areas that create emotions consist largely of evolutionary old systems; whereas our reasoning centers are located in the comparatively more modern cerebral cortex, which sits like a helmet atop the more ancient brain structures. The evolutionarily old brain areas, which regulate emotions, are foundational for vertebrate life. One of the earliest lessons imparted by evolution is that, to survive, animals must be attracted to what is good and wary of what is bad for them.

Extreme heights are dangerous places. Through self-controlled locomotion experience—crawling or having an infant walker—babies learn that heights, such as the visual cliff, are dangerous and so become wary of them. As we grow into adulthood, these feelings acquire higher stakes. As we'll get into later in the chapter, people who are afraid of heights will see heights as higher when standing at the edge of the cliff compared with people who are less afraid. You do not see heights as they are but as heights as seen by you.

Jerry Clore, a colleague of Denny's at the University of Virginia, has developed a theory of emotion linking how we feel with what we see and how we think. Essentially, negative emotions (sadness, fear) tell you to stop doing what you're doing—*Get away from the abyss!*—while positive emotions urge you to keep doing what you're doing—*Now that you're back on safe ground and feeling good, stay there.* They are imperatives: do, don't. Green light, red light.

The brain areas that create emotions include subcortical structures and the limbic system. Our limbic system and brain stem are responsible for regulating the body's internal environment of organs, glands, and metabolic processes. Affective and emotional feelings are typically accompanied by an awareness of the body's internal environment. For example, anxiety may entail an awareness of elevated heart rate, which is controlled by these ancient brain structures. Conscious awareness is a function of the cerebral cortex; and the cortex becomes aware of heart rate via sensors in the heart that transmit its state via the vagus nerve, a neural highway of sorts that connects the heart and other organs to the brain. The state of the internal environment, which is regulated by these ancient structures, is interpreted by the cortex as an evaluation of life's affairs.

The many reactions to the VR abyss from hell demo highlight the distinct roles played by the emotional and reasoning areas of the brain. The members of Congress experienced two forms of awareness running in parallel: the reasoning centers knew that the height was not real, while the emotional regulation centers detected danger and created a distress arousal. For congresspersons staring into an abyss, their anxiety likely entailed the telltale symptoms of a stressed-out emotional body: sweating, trembling, and an elevated heart rate. Sensors in the heart, skin, and muscles note the body's distress state and send that news up to the cortex. The congresspersons' anxiety, along with just about any other emotional experience that humans have, was the product of the emotional brain communicating with the neocortex by way of the body.

This runs against conventional, caricatured portrayals of emotion. Prototypically, the 2015 Disney animated movie, *Inside Out,* depicts emotions as eight distinct internal characters rising to the stage of awareness when properly cued. To Clore, this portrayal of emotions as distinct personalities is not at all an accurate description of our emotional world. Emotions are not something that we have but rather something that we create. Emotions are the result of our evaluating how good or bad a given situation is for us. "Sadness, fear, and anger are all negative affective reactions, but in each, the affect is occasioned by something different," Clore notes.[7] "Sadness concerns past bad outcomes, fear is about possible bad outcomes, and anger is about blameworthy actions of others causing bad outcomes. Emotions, therefore, differ from each other primarily because they are reactions to different kinds of situations." Emotions are not sprites entering into the mind from some Platonic casting room but rather responses to the ecologies within which we find ourselves. Emotions allow us to perceive good and ill in a world filled with both. They may seem to come unbidden, but they are, in fact, of our own creation, and they possess both the wisdom and vicissitudes of our oldest and most fundamental brain structures. Just as our bodily size, fitness, and the like frame the way we perceive our worlds, so do our emotional states—a matter of fact to anyone who's felt like they can't get out of bed in the morning.

DO, DON'T

WHEN JULIUS CAESAR and Augustus returned home victorious after battle, the Roman senate conferred on them the title *imperātor*, meaning "the person giving orders," closely related to *emperor*. Thus, the *imperative* mood in grammar, which "expresses the will to influence the behavior of another."[8] It is a staple of parenting ("Tie your shoes!"), advertising ("Just do it!"), and instruction manuals ("Turn the engine off!"). This is the tone of telling, pushing, exhorting, and commanding. Power in its elemental form. It is also what our feelings do to us: guiding action in many cases, commanding it in others.

The emperor of all feelings is pain. Its job is to protect the body from harm—*Take your hand off the stove. Now!* In his book *What the Body Commands: The Imperative Theory of Pain*, the Australian National University philosopher Colin Klein considers an injured ankle: "It aches. That aching keeps me from walking on it. I'm not sure why it aches. Maybe something is still torn down there, maybe it's just weak and needs special care. I don't know. Nor does it matter—whatever the reason, I should not walk on my ankle. If I can manage that, whatever is wrong will sort itself out. If I do walk on my ankle, it would just make things worse, maybe permanently. Rather than leave it up to me, my body just motivates me not to walk around if I don't have to. Then I will heal. That is the function of pain."[9]

As the emperor, pain is capable of the most urgent of all sensations: *Make it stop!* Rare cases of congenital insensitivity to pain (CIP) illustrate how deeply essential pain is to our lives. The absence of pain is a loss of the *imperative,* not the sensations of touch. Children with CIP have to be attentively cared for, or they will unintentionally harm themselves.[10] These children will burn themselves by placing their hand on a hot stove and leaving it there. They feel the heat but not the pain with its reflexive imperative, *Move your hand now!* Most of these children suffer from severe orthopedic problems, the reason being that they do not constantly shift their body posture to take the stress off their joints. When you or I stand, sit, or lie down, we are forever changing our position, rocking back and forth, crossing and uncrossing our legs, rolling over. These movements distribute the pressure on our joints so that damage is avoided. We do this because we feel slightly uncomfortable in our current posture and are thus motivated to change position. To be forcibly held in one posture is a form of torture. Children with CIP do not shift their postures because they are incapable of feeling uncomfortable; and as a result, they wear their joints out.

Psychologists tend to split the experience of pain into two interrelated processes, *pain sensation* and *pain affect.* The sensation is information about tissue damage, gathered and communicated by pain receptors, sent through the spinal cord up to the brain.[11] *Pain affect* is the bad, unpleasant feelings that come along with pain sensation, plus the emotions stirred up by impending tissue damage. Affect is where the imperative action is, since it's the affective experiences that signal that something is aversive, painful, and to be avoided, and thereby, provides the impulse to escape from or reduce whatever stimulation is bringing about all the unpleasantness. But, importantly, you can have the affect without the sensation—painful feelings happen without tissue damage.[12] Bones break. So do hearts.

SOCIAL PAIN

UNLIKE APPS THAT CAN be downloaded to your smartphone or computer, evolution must create new functions by modifying and building upon an organism's existing anatomy. For example, evolution co-opted small reptilian jawbones for new purposes during the evolution of the mammalian ear. Your middle ear has three tiny bones—the ossicles—two of which evolved from bones in our ancestral reptile's jaw. Similarly, selection pressures on our evolving need for social affiliation co-opted the emperor of all feelings to exhort us to "fit in." Not fitting in—being rejected and unloved—hurts with a pain as physically real as a punch in the chest.

The concept of "social pain" tracks back to Jaak Panksepp, an Estonian American psychologist who coined the term "affective neuroscience." Like so many prominent psychologists, Panksepp was drawn to the field as a route to understanding war, as his family fled Estonia during the Soviet invasion. In the late 1970s, he began doing nonhuman animal experiments to investigate emotions. In what would become a landmark paper, he and his colleagues gave morphine to puppies who had been separated from their mother and siblings. The opiate led to drastic reductions in crying as well as bodily distress signals like fidgeting. From this, he and his collaborators hypothesized that the neural circuits responsible for separation distress were an evolutionary elaboration on the circuits originally purposed for the perception of pain.[13] The pain of exclusion, they reasoned, drew from the same mechanics in the brain as physical pain, and so a palliative effective for one would work for the other. While controversial at the time, "social pain" has become a widely accepted theory within the social and brain sciences.

Human brain imaging research on social pain has made productive use of a mischievous video game called Cyberball. Playing this game is simple enough: the participant sees two cartoon avatars for fellow players, who either pass a ball between themselves or to the participant's avatar. (There aren't really players behind those other avatars; the whole thing is rigged in advance—but the "other players" provide a cover story to enable the experiment.) In the experiment in question, researchers created both *implicit exclusion* and *explicit exclusion* scenarios. The implicit condition occurred during the first brain imaging scan. The participant was told that because of technical difficulties, the scanner they were in couldn't connect with the scanners of the other players, so the participant could only watch the passing between the silent play partners. In the second brain scan, the *inclusion condition,* the participant was included in the ball passing. In the *explicit exclusion* scenario, the third scan, the participants received 7 throws—and then their partners spent the rest of the time, a full 50 throws, passing to each other.[14] The participants' brains were scanned the entire time, and they also filled out a questionnaire. The results, in line with the early canine studies, found that regions of the brain associated with physical pain were more active during the implicit and explicit exclusion—where participants reported feeling excluded and ignored—than during the inclusion scan. Brain areas previously linked to pain distress showed up with the distress of social exclusion—which also happen to light up when a mother hears a baby's cry. And in a remarkable parallel, a brain area associated with regulating physical pain distress was also active as the participant recovered from the stress of social exclusion.

Why would exclusion be painful? What's adaptive about experiencing social pain? As we'll explore in greater detail in the last third of the book, humans are profoundly social animals. In people and in other apes, social integration is a key to survival. This is well-documented within primatology: for example, even when controlling for how socially dominant a baboon mother is, babies of females who have better social connections have a better chance of surviving to their first birthday.[15] Hence the need for social pain. In the same way that the aversiveness of the cold provides an imperative to put on a coat, the aversiveness of social isolation prompts the prosocial behaviors that bring an individual, be they baboon or human, into the fold of the group. Social pain pushes people to avoid situations where they might feel their sense of inclusion threatened, hence why (most) people are quick to cease doing an offensive behavior once they're made aware of it, allowing them to keep themselves included. When you consider the group or clan as a whole, social pain is a way of establishing and regulating norms of behavior. One can also see why, throughout history, exclusion has been such a punitive tool, whether in the walled exclusion of prison or the outside-the-city-walls exclusion of exile.

OF FEELING AND ATTRIBUTION

IN THE SPRING OF 1981, Jerry Clore and his collaborators selected 93 telephone numbers from the University of Illinois at Urbana-Champaign student directory, calling them on either sunny or rainy weekdays in April or May.[16] The interviewers would introduce themselves as calling from a psychology department in Chicago—so as to indicate they were ringing from out of town—and then they continued the conversation under one of three scripts. First was an *indirect-priming condition,* where the interviewer would make a seemingly irrelevant aside of "By the way, how's the weather down there?" before steering the conversation again by saying, "Well, let's get back to the research. What we are interested in is people's moods. We randomly dial numbers to get a representative sample. Could you just answer four brief questions?"[17] Or, in the *direct priming condition,* the interviewer would say hello, mention they were calling from Chicago, and then plainly state that the researchers are interested in how weather affects a person's mood. Then, in a *no-priming condition,* the interviewer avoided any inquiries into the weather. Then, following whichever introduction, the interviewer asked all participants a series of four questions—asking students to rate how happy they felt about their lives as a whole on a one to ten scale, how much they wanted to change their lives, how satisfied they were with life as a whole, and how happy they felt in the particular moment.

After crunching the numbers, it became clear that subjects felt greater "momentary happiness" on the sunny rather than rainy days. This is not a surprise. In turn, the higher the momentary happiness, the greater the over-all life satisfaction and a sense that they needed to change less in their lives. Again, it is not surprising that if things are going well now, then you are likely to think that things are pretty peachy overall. The link between happiness and sunniness was strongest in respondents who hadn't been primed at all to think about the weather. For those participants who had been asked about the weather—directly or indirectly—rain or shine, the weather had no effect. In other words, if you are asked about the current weather, then it does not influence how happy you feel, whereas if you are not asked about the weather, then your sunny disposition depends on it being sunny and rainy days make you feel gloomy. According to Clore's research, if we don't know the source of what's making us feel a certain way, then we attribute our current affective state to ourselves.

Some of the most debilitating feelings accompany depression, which may be defined as an overall ennui or anhedonia—the inability to feel plea-sure even when engaging in what should be pleasurable activities. The World Health Organization says depression is the number one cause of disability worldwide, affecting 322 million people, or about equivalent to the entire US population.[18]

One innovative method (among many) for treating depression is concrete-ness therapy, which seeks to help people attribute their negative feelings to situations rather than to themselves. Working together, the counselor and counselee attempt to uncover what the depression is *about*, what it's responding to or stemming from. In the language of Clore, a goal of this approach is to transition from a "diffuse mood state to a specific emotion."[19] Instead of trying to avoid the feelings, called "emotion-focused coping" in the literature, you can address the issues themselves, called "problem-focused coping." The idea is to get people to find reasons for their negative mood in something else—the weather in Clore's study or a failed romantic relationship—rather than internalizing responsibility for all their despondent feelings.

One of the cognitive habits associated with depression is rumination—a tendency of the mind to keep returning to an object of worry, where you recurrently ponder about the causes and meanings of why you feel bad.[20] Multiple longitudinal studies indicate that one's tendency to ruminate predicts how likely and how intense depression will be.[21] This contributes to generalization, another key factor in depression, where single events are wantonly amplified into universal principles. Perform poorly on an algebra test, and you're bad at math; get into a conflict with your teenage daughter, and you're a bad parent. Again, the emotional regions of the brain are more than capable of putting us in a positive or negative mood, but they are not adept at assigning causes for these feelings. It might be the weather or it might be you.

Ed Watkins, a clinical psychologist at the University of Exeter, has tested concreteness interventions in the lab and in clinical populations. To prime abstract thinking, in one experiment, the researchers asked participants to imagine fictional social situations, like an argument with a close friend or a successful job interview.[22] The participants were then asked to analyze the event in one of two ways. They were asked to think in terms of *why* the event unfolded as it did and make sense of its meanings, causes, implications, and the like. Or they were asked to focus on the mechanics of the event itself: how it happened, imagining the "movie" of the exchange as it unfolded. The first prime prompted the kind of abstract thinking associated with rumination, while the second prime prompted concrete observations about how and what happened. In line with this experimental research, Watkins has also been administering this concreteness therapy in clinical settings. He's found that it's helped people with depression describe their problems with greater concrete detail *and* also reduced levels of self-criticism, rumination, and the risk of relapse into another depressive episode. While more clinical and experimental research needs to be done and there's no one-size-fits-all solution to well-being, his work shows how attribution itself can be an avenue for encouraging mental-emotional health. When Drake spoke with Watkins, he noted that many of the clients in his clinical practice are parents. He'll coach them to think more specifically about how and what happens when, for example, they have conflict with their kids. If you get into an argument with your teenager, it can be easy to spiral into negative, self-aggressive, recurrent thoughts: *Why am I a bad parent? Why do we fight so much? What's wrong with our family?* But if you try to play back the "movie" of the situation in your head, and think about the teen's perspective and yours and everything that was going on that day and whether or not you're sleep deprived or underfed or overstressed, and you told them to do their homework and they ignored you and you got angry, then maybe taking it step by step will allow you to see what could be done differently, like taking a more gentle approach to the conversation. "That's grounding it in the context," Watkins says. "If I changed a bit of my behavior, then it wouldn't have got so heated. That would move it away from the sense of 'I'm just a bad mom'—really grounding it in the particular instance that happened."[23]

The red-light/green-light signals that emotions provide tell us that if we feel good, then we should keep doing what we are doing, but if we feel bad, then we should stop. Yet being told to stop does not tell us what to do instead. This is a dilemma of depression that often results in withdrawal and freezing responses. It is hard to know what to do when your mood tells you to stop what you are doing yet offers no alternatives. The work of Watkins and other clinicians suggests that depressive people are at risk for seeing some aspect of their personalities or makeup as the reason why something bad happened; and it's helpful to learn to challenge that universal-style thinking and instead examine the specifics of a situation as a supplement to other treatment.

The slipperiness of attribution matters not just for mental health but for morality. In another Clore study, participants were brought into a deliberately disgusting experimental testing room—the only one available, they were told, due to space limitations—and asked to take a seat at a table.[24] The table had become sticky with rotting food and was situated next to a trash can overflowing with old pizza boxes. To top it off, after an experimental confederate apologized for the state of the room, they handed the participant a chewed-up pencil. This pencil was to be used to complete a questionnaire consisting of a number of ethical quandaries. A control group of participants came into the same room after it had been wiped clean of refuse. One of the ethical problems was a classic thought experiment, called the trolley problem, which was invented by the British philosopher Philippa Foot in 1967.[25] The scenario—which is familiar to many an undergraduate—goes like this. Imagine you're standing on a street, and you see a runaway tram hurtling down a track on which five workmen are toiling away, unaware of the out-of-control vehicle that imperils them. But you notice that a switch will cause the track to branch off to another side, where just a single worker is toiling. "How wrong is it for you to hit the switch that would cause the trolley to take the track on which one workman stood in order to avoid the deaths of the other five men?" the participants were asked. It's not an easy decision. In general, most people say that they would flip the switch, causing the unfortunate loner to die while the other five lived.

Clore and his colleagues also asked their participants to take a standard personality test used to assess how much an individual paid attention to their own bodily physical states. They found that the people who habitually paid attention to their bodies, and who were thrust into the disgusting room condition, judged the questionable moral action of yanking the lever of the trolley track to be more unethical than people who sat in a clean room or who didn't pay much attention to interoception. For the people with greater body awareness, the repugnance of the space colored the repugnance of the moral dilemma.

EMOTION SHAPES THE SIZE AND SCOPE OF PEOPLE AND THINGS AROUND YOU

IN 1675, FATHER LOUIS HENNEPIN, a Flemish priest by training and missionary by trade, was dispatched by his order, following the wishes of Louis XIV, to survey the interior of New France (now Canada). After spending a couple years serving churches around French settlements, he set out to explore the interior. In doing so, he became the first European to document Niagara Falls. Viewing the waterfalls from above, the friar estimated the height of the plunge to be 600 feet, but the actual height is just 167 feet. He noted how frightening the sight was: "The two brinks of it are so prodigious high, that it would make one tremble to look steadily upon the water," he wrote in his journal.[26]

Some 334 years later, Jeanine K. Stefanucci—a graduate student of Denny's, who is now a faculty member at the University of Utah—followed up on the peripatetic friar. "Did arousal associated with a fear of the height influence his estimate of how high the falls were?" she asked in a paper. In four experiments, she and a collaborator found that when participants had high emotional arousal—when they were excited or frightened—heights appeared higher, similar to the priest's early colonial self-report.[27] The experiment's participants stood at the top of a balcony. The two floors below them seemed farther away if they'd just seen a range of emotionally charged images than if they'd seen neutral ones. But if they down-regulated their arousal—by deep breathing—the overestimation effect lessened. Emotional arousal shapes the way we see, think, and feel. Elizabeth Phelps, a psychologist at NYU, has shown that when people are frightened, their vision becomes more sensitive to contrasts—meaning your visual acuity improves such that you can better distinguish between subtle shades of gray,[28] suggesting that being afraid makes it easier to spot threats around you.

In that experiment with the balcony, as well as those with steep hills, researchers have consistently found that people at the top of a given precipice will report greater estimations of distances compared to people at the bottom. There are intuitive explanations for such a thing: you can tumble down a mountain to your death, but not up it. In a related experiment, Jeanine, Jerry Clore, Denny, and UVA undergraduate Nazish Parekh further investigated this link between fear and perception, with the help of a wooden box and a skateboard, both perched at the top of a steep hill. Participants, none of whom had experience with skateboarding, stood on the box or the skateboard and made the same set of estimations as in earlier slant-perception tasks: a verbal estimate, visually matching shapes, and matching a tiltboard to the slant. And as with earlier studies, the verbal and visual were greatly overestimated, compared to the relatively accurate tiltboard. After making the slant estimates, participants rated their fear of descending the hill on a 6-point scale. Among the people who stood on the skateboard, feelings of fear were associated with greater overestimations of the hill's steepness compared with those who were less afraid.[29] These findings buttress case-study reports from clinical psychologists. In one stunning example, a patient with acrophobia—extreme fear of heights—"reported that as he drove towards bridges, they appeared to be sloping at a dangerous angle."[30]

While it took some effort, Clore convinced Denny that emotion may influence hill-slant perception, which prompted this branch of his vision research. In one experiment they worked with one of Denny's graduate students, Cedar Riener, now a faculty member at Randolph-Macon College, and they asked participants to listen to either happy (Mozart's *Eine kleine*

Nachtmusik) or sad (Mahler's Adagietto) music while looking at a hill.[31] Sure enough, the participants who had their heartstrings pulled by the melancholy Mahler music saw hills as steeper than those who had listened to Mozart's cheerful music. Intriguingly, the overestimation was of a similar magnitude to that of participants, who in earlier experiments, had been asked to lug backpacks full of weights. (People experience "sadness as a burden," Clore would quip in a later paper.[32]) There's a deep logic to this: feeling sad is relevant information about the resources you have available to ascend a hill. If you're depressed, it's going to be more challenging to climb a hill than if you're feeling fit and energetic.

Perhaps the most hair-raising strain of research in the field comes in the form of work being done with spiders—and how people perceive them. Similar to the studies on fear of heights, people with higher scores on tests measuring how afraid they are of spiders tend to see the eight-legged creatures as larger,[33] and the more scared they are of them, the closer they'll estimate a live tarantula to be when standing across the room from it. Jessi Witt, Denny's protégé and collaborator, found a fascinating overlap between threat and ability, relative to things that frighten us. She asked participants to sit at a table with a video game of sorts, where an object like a spider or ladybug would appear. It would scurry around and move toward them, and they had a paddle of various lengths that they could protect themselves with. The result: the spider looked like it was moving faster than the ladybug—especially if participants could only defend themselves with a tiny paddle. Perception expresses the relationship between you and what's around you, so in our own personal perceptual world, the things you fear will, by extension, loom large.[34] Crucially, what we fear is not only cliffs and spiders but also each other.

Some of the tarantula researchers carried out perceptual experiments with a paid male actor in place of an arachnid.[35] In the *threat condition*, they watched a video of the guy talking about how his favorite hobby was hunting, how much he loves the feeling of holding a gun, and how he felt that in the city there was no real way of getting his aggression out, so much so that he felt like he was going to explode. In the *disgust condition,* the same young man talked about when he had a summer job at a fast food restaurant and did various repugnant things, like spitting in customers food or urinating in their soft drinks. In the *neutral condition,* the same guy just talked about the classes he was slated to take in the next semester, speaking in an even-keeled manner. The participants were then brought into a room with that same male actor, where they had their heart rate measured and answered surveys on how threatening or disgusting the actor was to them. Finally, they also estimated how many inches they were from this person, who sat 132 inches (11 feet) away. Averaging the scores, the actor appeared much closer to participants when they'd been exposed to the threatening video versus the disgusting or neutral ones—55 inches verses 78 or 74 inches, respectively.

Again, in the uncompromising light of evolution, the job of perception is not to represent an objectively accurate world, but to provide information in a form that will promote survival. Emotions, Clore shows us, have the job of giving a red or green light toward approaching or avoiding objects, people, and situations, and shape perception accordingly. Regrettably, our negative prejudices can color our perceptions of stigmatized people in ways that result in disastrous outcomes. This has been studied by Stanford social psychologist Jennifer Eberhardt, who has won a MacArthur genius grant for her work.[36] In one experiment of hers, white male university students watched one of two videos: one showing entirely African American male faces in quick succession, another showing male faces of multiple races. Then the researchers placed a blurry object on a screen that looked like television static. Across a brief duration, that object slowly came into focus. The object being made visible was either crime related, like a gun or a knife, or crime irrelevant, like a stapler. Participants who had just seen the reel of black faces more quickly identified the threatening but not the neutral objects. "It's almost as though because blackness is so associated with crime that you're ready to pick out these crime objects out of the environment in a way that you're not if you're exposed to a white face," she said in an interview.[37] In another experiment,[38] Eberhardt and colleagues asked Stanford undergrads to rate how stereotypically "black" people in a range of photographs looked—all of which, unbeknownst to the participants, were pictures of people who had been convicted of capital crimes and were eligible for the death penalty. She and her colleagues then compared how stereotypical were the judgments given by Stanford students with the severity of the sentences handed down by our legal system. She found that those people with the most stereotypically black faces were twice as likely to receive a death sentence. The more black you appear, the harsher the sentence. In another experiment, conducted by Joshua Correll and colleagues at the University of Chicago, participants playing a video game were asked to identify individuals holding a threatening or neutral object—a wallet or a gun. If they're holding guns, then participants were instructed to press a button labeled "shoot" and if it's not a gun, then they were to press another button labeled "don't shoot." It was found that participants will respond with "shoot" faster to an image of a black person with a gun than a white person with a gun, and they're more likely to mistakenly click "shoot" for a black person with a neutral object than a white person without a gun.[39] Importantly, for this last study, this bias was moderated by whether the stereotypical link between black people and "danger" had been reinforced by reading a news article about blacks and violent crime. Again, you do not see the objective world, whether of objects or people, but the world as seen by you, and all the emotional associations you've accumulated over a lifetime. And as we'll explore in the next chapter (Speaking), as well as in the last part of the book (Belonging), much of our experience is of a social world.

6.

SPEAKING

HANDS, MOUTHS, AND MIMICRY

MOST THEORIES OF HOW LANGUAGE evolved assume a foundational role for the movements of our hands.[1] Communicative gestures are thought to have preceded or coevolved with spoken language, with speech and communicative gestures both making use of common brain systems that first evolved for controlling our hands as we reach and grasp for things in the world. Functional neuroanatomy attests to the link. In most people, the left hemisphere of the brain is associated with both language and skilled gestural movements.* Indeed, it is common neurological knowledge that lesions in the left hemisphere often lead to the co-occurrence of both aphasia (an impaired ability to speak or listen) and apraxia (having difficulty with familiar fine motor movements like tying shoelaces or buttoning a shirt).[2] Activity in Broca's area, a region related to speech production, also appears to be involved in manual reaching and grasping and also in simply *watching* other people use their hands.

The common ancestor for all primates appeared about 65 million years ago. This protoprimate looked a lot like a squirrel and probably scurried about in trees much as squirrels do today. Now if you're a squirrel and want to carry a nut home, then you will carry it in your mouth. Just like any other quadruped, your mouth is the only means you've got for holding and carrying your stuff as you move about. We humans, of course, have hands with which to carry our baggage; but the manner of carrying, which is shared by the mouth and hands, is an evolutionary connection that persists. It is also evident in our speech.

If you are a human carrying baggage in both hands, then what do you do if you need to carry something else with you? You will carry it in your mouth if you can. If the object is big—like an acorn for a squirrel or an apple for you—then the mouth needs to be wide open; whereas if the object is small, like a pencil, then the object will be carried with a closed mouth that holds the object with the lips or teeth. Similar gripping postures apply to our hands. An apple is held with an open grip—apple pressed into the palm with the fingers. This is called a "power grip." The pencil is held with a closed grip, called a "precision grip," in which the pencil is grasped between the thumb and index finger. As this example makes evident, there is a "mouth-

---

* This is true for almost all right-handed people, but somewhat less so for left-handers. The left hand is always controlled by the right hemisphere; however, most left-handers perform language processing in the left hemisphere.

hand parallelism." Big things are carried with an open posture and small things with a closed posture in both the mouth and the hand.

A Finnish research team, Lari Vainio and colleagues, conducted an ingenious study to assess whether this mouth-hand parallelism is evident in the speech sounds that we make when describing small and large objects.[3] Their experimental paradigm follows this simple logic: if two behaviors make incompatible demands on the same underlying neural system, then performing one of these behaviors will interfere with the simultaneous performance of the other. That you can easily walk and talk at the same time implies that walking and talking do not rely heavily on the same brain areas, whereas speaking and listening to speech is really hard because speaking and listening do rely on many of the same brain areas. Try to silently count backward by 3s from 181 while also silently reciting the alphabet. Or try keeping up a conversation with a friend while reading an article on your phone. Can't be done. Both tasks demand the full resources of the language areas of the brain.

In Vainio's study, participants held a device that had two switches; one switch could be activated when pressed with a closed precision grip and the other with an open power grip. To picture this device, imagine that you are holding the handle of a hammer. Affixed to this handle are two switches that are equally easy to engage. One is activated by squeezing the handle—this is the open power grip. The other is a small switch at the top of the handle, which is depressed by pinching the thumb and index finger together—this is the closed precision grip. Participants watched a screen on which either a blue or a green stimulus appeared. If blue, then they were to press the precision grip switch, and if green, the power grip switch. In both cases they were to respond as fast as they could. Easy-peasy. Reaction times were collected and the association of color and grip type was counterbalanced across participants. At the same time as they were making these grip responses, they vocalized simple consonant-vowel syllables that simultaneously appeared on the screen. You can try it for yourself with example syllables:

ti

pu

Notice how in both cases your mouth takes on a closed posture, one in which you could hold a pencil in your mouth. Now say these two syllables:

ka

ma

Notice now how your mouth takes on an open posture appropriate for holding a small ball or apple. So, to review the design, participants were required to simultaneously respond to a reaction-time task with either a closed (pinch) or open (squeeze) grip while simultaneously vocalizing a syllable with either a closed (*t*) or open (*ka*) mouth posture. You may have already guessed the results. If both the grip and mouth postures are compatible (both closed or open), then responding to the reaction time task was faster than if they were incompatible—a closed hand and open mouth or vice versa. Clearly, the distinct behaviors of controlling the hand and articulating speech are drawing on common neural systems. If both are doing the same thing, then they interfere with each other less than when they are attempting to make incompatible movements. It's easier for our hands and mouths to move with similar than with dissimilar movements, which attests to a parallel between how our hands, mouths, and even language evolved. In many cases, the sounds for words mirror how objects are held.

By this account, it is not a coincidence that *little, tiny,* and *petite* are adjectives meaning "small" and are pronounced with a closed mouth, or that *large, huge,* and *humongous* are articulated with open mouths. The Berkeley linguist John Ohala has noted this trend across languages: English has *teeny* and *wee* for "small," and *humungous* for "large." Spanish has *chico* versus *gordo*. French *petit* versus *grand*.[4] Greek *mikros* versus *makros*. Japanese *chiisai* versus *ookii*.

Thus we see that the evolution of grasping went from the *mouth* to the *hands* and back again to *speech*. Whether or not you have to use an open or closed grasp to hold a given object has been preserved for millions of years, tracing back to our protoprimate ancestor that walked on four legs and carried things in its mouth. The "huge" acorn that required an open mouth to hold 65 million years ago now requires an open mouth to describe its size.

This controversial argument goes against the long-standing traditional view that the sounds of words and the meanings they represent are disconnected and arbitrary, that the signifier—the word and how it sounds—has nothing to do with the signified—the object or concept it's referring to. This assumption about the arbitrariness of sign and thing signified traces back more than century ago to Ferdinand de Saussure and the birth of structural linguistics. In opposition to this view, "sound symbolism" or "phonological iconicity" are modern ways of describing the phenomenon wherein words sound like what they mean. These icons are often hidden in plain sight, like the nasal sound *sn-*, which is often at the beginning of schnozz-related words, like *sniff, snore,* and someone who walks around with their nose in the air, a *snob*.[5] There are also *phonesthemes,* where particular sounds tend to pair with particular things or feelings or concepts, like *gl-*, which manages to *glimmer, glitter, glisten, glow,* and *gloom*.[6] The list goes on.

The evolution of dexterous hands was made possible by our ancestors' transition to bipedal walking. With bipedalism, objects that had been carried in the mouth could now be carried with the hands, and thereby the carrying function of the mouth transitioned to the hands. During development, infants and toddlers first explore the world with their mouths and then their hands, and what they discover about the world are its many affordances for mouth-ing and hand-ling. For infants, the mouth and hands are complementary means for discovering the properties of the world and they continue to cooperate as language emerges and develops. We talk, not only with our mouths but also with our hands.

TALKING WITH YOUR HANDS

GESTURE IS SPECIAL. It's not sign language, where each individual hand configuration and motion is a word unto itself, assembled together in sequence like beads on a string. Gestures are something entirely different. To Susan Goldin-Meadow, a psychologist at the University of Chicago, gesture is part of how we experience, express, and understand language.

In the late 1990s, Goldin-Meadow and her colleagues set out to investigate why people gesture. Two hypotheses motivated the research: (1) it could be that people gesture because they see other people doing so, or (2) it could be that they gesture in order to help listeners understand what they're trying to say. Of course, it could turn out to be something else entirely. To answer this question, a handful of congenitally blind children were recruited to participate in a study conducted at the University of Chicago.[7]

The blind participants were all between 9 and 18 years old and had no other physical or cognitive issues. They and a group of sighted peers, matched for demographic backgrounds, were videotaped while responding to a series of prompts found to elicit gesturing in sighted children. These prompts were taken from an earlier study by Goldin-Meadow and her colleagues in which they were assessing whether children understood that a liquid's quantity stayed the same when poured from one glass to another. (Young children tend to get this problem wrong because they are seduced by the liquid's different appearances in glasses of different shapes.) In that earlier study, the children were asked to explain their answers, and they said things like "If you pour the water back into the first glass then it will appear the same as before."[8] As they did so, they would often pantomime pouring water from one glass into the other. With that precedent in mind, the same questions were given to the blind and sighted children, who were also videotaped as they provided their answers. Coders then examined these tapes to evaluate the gestures of the two groups.

Both groups of children gestured as they spoke, and the blind children did so at about the same rate as their sighted peers. What's more, the kids all made similar movements when talking about similar things. Like, for example, when talking about pouring liquid from one container to another, both the sighted and blind participants held a C-shaped hand in the air, pantomiming holding a glass, then tilted it as though they were pouring water from one glass to the next. This suggested that people don't need to see someone else gesture to do the same themselves. Gesturing is not necessarily learned via observation, so the first hypothesis was rejected.

Then, in a follow-up experiment, another set of blind children were asked to complete the same task, explaining what was going on to an experimenter who they were told was also blind. In this experiment, the blind children gestured when talking to a "blind" experimenter just the same as when they were speaking to a sighted one. Therefore, gesturing is not necessarily performed for the benefit of listeners, so hypothesis two was rejected.

"Our findings underscore the robustness of gesture in talk," Goldin-Meadow concluded. "Gesture does not depend on either a model or an observer, and thus appears to be integral to the speaking process itself. These findings leave open the possibility that the gestures that accompany speech may reflect, or even facilitate, thinking that underlies speaking."[9] So if you don't learn to gesture from seeing other people do it, and it's not necessarily about communicating what you're trying to say, where does gesturing come from, and what is it for? Why are we compelled to make immaterial gestures whenever we speak? Goldin-Meadow, who has spent decades investigating this question, shared her thoughts with us. "Gesticulation isn't divorced from speech. It's completely tied to your speech. It's part of your cognition. It's not just mindless hand-waving."[10] While the exact mechanisms haven't yet been precisely mapped, she thinks that gesturing helps with the "cognitive load" of speech. We talk with our hands because it helps us with what we're trying to say.

Goldin-Meadow is careful to distinguish gesture from other forms of movements. It's not a pragmatic action like tying a shoe, cracking an egg, or throwing a ball. It is not dance or exercise, movements for the sake of movement that are a way of reaching some larger end, be it self-expression, performing art, or getting fit. Rather, gestures are "representational actions that do not have a direct impact on the physical world," she has observed, "but impact the world indirectly through their communicative potential."[11] Yet gestures are also different from what people who study these things call *emblems,* or different hand-shapes that have an agreed-upon, cultural specific meaning, like by forming a circle with thumb and forefinger to tell a friend that they're okay, or extending the middle finger to tell an antagonist the opposite. "If I say 'okay,' I can't use my middle finger," she says.[12] "But if I say something goes up, I can use my whole hand, my thumb, a finger, there isn't one way to do it. But with emblems, there is one way, it's like a word, in a way. When you make the shush sound, there's only one way to do it. When you're going scuba diving, you agree before you go down what a hand shape means."

Whereas speech follows the syntactic rules of a language, gesture is mimetic—it tries to imitate the thing that it represents.[13] (Hence the *miming* partygoers engage in during a game of charades.) In doing so, gestures communicate certain kinds of information in a manner that words alone can't easily convey. Imagine that you are telling someone about holding a newborn baby. Your arms will likely fold themselves around the imaginary infant as you press it gently and securely to your chest. There is emotion, tenderness, and protection in your gestures. Hands and arms were made to hold babies; and gesturing as such communicates this primal loving act with an immediacy that the speaking voice struggles to achieve.[14] This all suggests that gestures are a way of communicating, or maybe even understanding, things that are fundamentally embodied, like holding a baby.

Goldin-Meadow and her colleagues have had to design some clever experiments to get at how and where gestures help us think. It starts early: in one intervention with toddlers, her research team taught one group to point at objects whose names they were trying to learn, while the other group was not. The kids taught to point gestured more when speaking with their caregivers, and, two weeks after the eight-week training, they had larger vocabularies than the others.[15] Primary schoolers instructed to gesture while working through equations also do a better job of applying lessons learned about solving equations to new problem sets.[16]

Gesture, and especially pointing, paves the way for acquiring language in the first place. Before children are able to speak, they're making gestures in the form of pointing at objects, sometime around 9 to 12 months.[17] It's about pointing and naming, and thus putting the "index" into index finger,[18] like pointing at a cup and saying cup.[19] Yet there are actually two kinds of pointing, and kids as young as one year old do both: imperative and declarative. Imperative pointing is when you (or a baby) are pointing at something because you want it, like a cupcake tantalizing you from across the table. Declarative pointing is more sophisticated: instead of *gimme that* it's *look at that*. These are among the first, if not the first, opportunities for a baby to initiate what's called "joint attention" with its caregivers—you're looking at the red plush toy, I'm looking at the red plush toy, and we both know that we're both looking at the red plush toy. That, in turn, allows for knowledge to be communicated: baby points at toy, mom shares that same attention, mom calls it "Elmo," and baby begins to attach the vocalized sounds to that particular object, and soon its category. Sharing of attention in this way is an especially human pattern of cooperation—nonhuman apes do imperative pointing but not the declarative kind.

The *index* finger gets its name from the Latin for "pointing out," *indicō*. Yet, we point, not only with our fingers but also with our eyes. There is no need to be forever pointing at one thing after another when talking sensibly to young children. Children will naturally look at the targets of another person's gaze. As is the case with pointing, other great apes don't pay much attention to the direction of eye gaze, but human infants do.[20] Moreover, the appearance of our eyes has evolved to make the direction of gaze especially obvious. First, we have a white sclera, which is the part of our visible eye that surrounds the colored iris. (The sclera is more commonly known as "the whites of your eyes.") All the other great apes have a brown sclera surrounding a brown iris, making gaze direction difficult to discern. Second, the eyelids and skin surrounding our eyes create a horizontal almond shape that shows large areas of sclera on either side of the iris. This horizontally elongated shape makes the left/right direction of gaze really easy for others to see. (Take a look at the woman's eye on the jacket of this book. Regardless of whether you see her facing forward or to the right, you can tell that she is looking at you.) The eyes of the other great apes have a much more rounded apparent shape and show very little sclera. Evolution has turned our hands and eyes into instruments of communication, such that when speaking with others, we use them both to direct our listeners' attention and to help them create the meanings we wish to convey.

INDEXING IS VERBAL, TOO

JUST AS POINTING IS an invitation to pay attention to and learn about a target object, for young infants, hearing a voice is also an invitation to learn about what is going on at the moment. Many experiments have used the following design to see the links between vocalizations and learning about what objects belong to what groups or categories. First, different images from the same category (food, animals, etc.) are successively presented to an infant, who is looking at a video screen. The images might all be of animals—dog, horse, elephant, and the like. After this initial familiarization phase, two images are shown on separate video screens, one of a new animal, a tiger for example, and the other of something from a different category, like a head of lettuce. The amount of time that the infant spends looking at each of these novel images is then recorded. If there is an overall significant difference in how long the infants look at the novel animal versus nonanimal, like tiger versus lettuce, then it is concluded that they perceive a difference between them based on one belonging to the familiar category, in this case animals, and the other belonging to some other category. (Without yet having the words, they're realizing that "this belongs to that.) The idea with this experimental design is to get infants familiar with images that belong to a certain category and then see how they react to new images, one within and one outside of the familiar category. (Pretesting or counterbalancing of images is needed to assure that results are not simply due to one image—a tiger—being more interesting than another—a head of lettuce.)

Alissa Ferry, Susan Hespos, and Sandra Waxman, then all at Northwestern University, conducted a study with a design like that described above, with three-month-old infants.[21] They used images of dinosaurs for the familiarization phase and fish for the noncategory test stimuli. They found that their infants showed no preference for the in- versus out-of-category images during testing. They happily look, for example, at the fish as much as the novel dinosaur, and thereby showed no evidence of category learning. But, if during the familiarization phase, the presentation of each image was accompanied by a human voice, then when tested, the infants looked longer at the image that belonged to the dinosaur category than to the image that did not, a fish. If the images during familiarization were accompanied by the scream of a lemur, a small primate, then infants also learned the category and preferred dinosaurs. Accompanying the familiarization images with tones or recordings of human voices played backward did not evoke category learning. The critical evocative stimulus for learning was a human vocalization or that of another primate. Like gestures, vocalizations are embodied social behaviors that encourage infants to pay attention, learn, and put things into categories.

In that first year of life, babies tune in more and more to human speech. By about 6 months of age, the evocative power of nonhuman primate vocalizations drops out and infants learn categories only in response to hearing human speech. By 12 months of age, infants make use of novel words that are paired consistently with the to-be-learned category. Consequently, 12-month-olds will not form categories if the familiarization images are accompanied by familiar words, as in "Look at that," but they will learn if each image is accompanied by "Look at the [blank]," where *blank* is the novel word, like *fauna,* that seemingly names the category.

EMBODIED ETYMOLOGIES AND THE GROUNDING PROBLEM OF MEANING

BUT WHERE DO WORDS get their meanings from in the first place? It certainly goes deeper than the dictionary. The symbol merry-go-round argument—a philosophical thought experiment—puts this into stark relief. Imagine you just landed in a foreign city, and upon exploring the airport, you realize that everything around you is written in a language you can't recognize. You have a dictionary with you that's in the local tongue, but no translations into any languages you speak. You can look up the definition, but you're only going to get more symbols you don't understand. Suppose, for example, that the airport is in Japan and you want to know the meaning of the Japanese word (spelled ideographically), タクシー. You look up the word in a Japanese dictionary. タクシー: 通常、走行距離によって決まる運

賃で乗客を運ぶ自動車. Not very helpful. (Here it is in English: "Taxicab: an automobile that carries passengers for a fare usually determined by the distance traveled."[22]) To put the argument more strongly, imagine trying to learn a *first* language—Japanese, Italian, English—with only a dictionary to tell you what anything means. It would be a merry-go-round of one symbol to the next, words referring to words without ever referring to actual stuff out there in the world. To get off the merry-go-round, you need to relate the symbols to things you know through experience. You need to index the word to a perceived object. Without someone first pointing out what the words mean, or through systematic vocalizations inviting you to discover for yourself their meaning, you would be at a loss to learn them.[23] Returning to the Japanese airport example, you need to discover that タクシー refers to one of the cars waiting at the airport exit that will take you to your hotel for a fee.

Learning word meanings for concrete objects seems simple enough. The meaning of *taxicab* can be understood by pointing at one. But what about abstract concepts, such as *time*? According to the Berkeley philosopher George Lakoff, we understand the meaning of abstract concepts by inferring that they are similar to concrete objects and events with which we are familiar. The abstract concept of *time,* for example, is presumed to be like *distance,* a concept with which we have ample experience. Once the metaphor is acquired, we talk about time using the vocabulary for distance: time is getting *short,* the politician's speech *went on and on,* the story took a *long* time to tell, and so forth. As another example, the abstract concept, *theory,* is often thought of as being like a *structure*: it is *well-founded* or has *shaky foundations,* is *all-encompassing* or *narrow,* will *collapse* or *hold up.* Similar metaphors are used across languages: heavy is a common signifier of importance in English (*weighty*) and German (*wichtig,* or "important," and *gewichtig,* or "weighty"). We draw our concepts from our experiences—the left-handed compliment, the right-hand man.

Language that refers to actions is understood, according to Lakoff, by simulating these actions ourselves. Research supports this notion. For example, when reading about transferring responsibilities, the muscles in our hands will make very subtle movements in much the same way as when reading about transferring objects, like putting away the dishes.[24] Brain imaging studies have found that the corresponding motor cortex areas for the foot, hand, or tongue will activate when you read about kicking, picking, or licking.[25] (The neuroscientists involved concluded that their findings ruled out that there's some unified "meaning center" in the brain that makes sense of all language, but rather that you use different parts of your brain depending on what a word means.) Follow-up work found that reading about tools activates brain areas related to hands, and food to the mouth.[26] The individual differences of *your* body also play a role: Daniel Casasanto, who carried out the bulky ski glove experiments that we profiled in the Grasping chapter, found that lefties and righties show opposite activations in the motor regions of their brain when reading action verbs, as in *grasp* or *throw*.[27] The meaning of *throw*—when reading "Please *throw* the ball to Taylor"—is understood by activating motor areas of the brain that enable you to throw, left hemisphere for righties, right hemisphere for lefties. More recent brain imaging research has also found that abstract words prompt motor system activation—simply reading words like *love, thought,* and *logic* lead to motor activations usually associated with the face.[28] The upshot of all this is that the evolution of language must have scaffolded itself on top of neural structures that were present in our prelinguistic ancestors. So of course the way we make sense of language is going to make use of what evolution has tuned us to do—the reaching, grasping, moving, feeling, and the like.

As a final observation about metaphors, consider: What are we talking about when we talk about a *grounding* problem anyway? What is meant by the assertion that meaning must be grounded in something other than the abstract symbols of the symbol merry-go-round? What is the ground to which we're referring? The earth of our experience is, of course, our perceptual worlds, our personal *Umwelt*. The grounding problem of language is solved when we connect what we hear and read to our store of embodied experiences.

WHAT BOTOX TELLS US ABOUT READING

IN A 2010 PAPER,[29] Arizona State psychologist Art Glenberg and his team reported a study showing a surprising consequence of getting Botox injections, the cosmetic dermatological treatment that earned more than $3 billion in revenue worldwide in 2017.[30] Botox is derived from the poisonous bacteria *Clostridium botulinum,* which can, in some cases, cause botulism, a disease that attacks the body's nerves and can cause paralysis.[31] There have also been promising results using the drug to treat migraines and facial spasms,[32] but it is popularly known for its youthful-appearance effects, smoothing out the fine lines and wrinkles that come to the face with age. It does so by effectively paralyzing facial muscles under the skin, including those involved in grimacing and frowning, like the forehead's *corrugator supercilii,* whose Latin name literally translates as "wrinkler of the eyebrows." Both before and after a Botox treatment, the researchers asked the participants to silently read and understand sentences of contrasting emotional qualities. Participants were quizzed on their comprehension to assure that they were reading carefully. Some sentences conveyed angry feelings ("Reeling from the fight with that stubborn bigot, you slam the car door"), some were happy ("Finally, you reach the summit of the tall mountain"), and some were sad ("You hold back your tears as you enter the funeral home"). It was found that following Botox treatments, users took *longer* to read the sad or angry sentences, but the drug had no effect on reading times for the happy ones.

Why would that be? In accord with Lakoff's account of how we understand actions, Glenberg proposes that reading comprehension is a matter of simulation: the brain simulates the experience being portrayed in the text by enacting it, at a small scale, through the body. So to understand or express sadness or anger, you have to furrow your brow, at least to some extent. Now, since it is difficult or impossible to furrow your brow when the muscles required to do so are paralyzed, it takes you longer to understand text about sadness or anger. Smiling does not engage the forehead muscles, and consequently reading happy sentences is unaffected by the Botox treatment. The motto for Glenberg's lab at Arizona State sums it up: *Ago Ergo Cogito*—"I act, therefore I think." To make meaning of language, we're constantly enacting, visualizing, and grasping at it.

Another study of Botox users found that they were slower to perceive subtle but not exaggerated emotions when viewing peoples' faces.[33] This presents a fascinating implication, that our embodied experience gives us a more fine-tuned sense of others. From an embodied perspective, we understand other people by putting ourselves in their shoes, or maybe eyebrows. While this perspective might feel intuitive in the confines of this book, it is in fact quite controversial. Similar to the computational theory of mind, language is often thought to be a collection of abstract rules and symbols and no more.

The embodied nature of metaphor has been revealed in studies using different forms of congruency between language and movement. In one study, participants were asked to make upward or downward hand movements while reading sentences pertaining to up and down.[34] Some of the sentence conveyed the literal meaning of up/down as in: "The pressured gas made the balloon rise." Others conveyed their metaphorical meaning: "His talent for politics made him rise to victory" A third set of sentences conveyed abstract meanings consistent with an up/down metaphor. For example, *up* is implied in "His working capability made him succeed as a professional." The hand motion either matched or mismatched the literal, metaphorical, or implied meaning of the sentences. It was found that hand motions were faster in all three conditions when they were consistent with the meaning of the sentence.

This finding shows that literal and metaphorical meanings engage neural systems that are also responsible for moving the hands up and down, similar to the mouth-hand parallelism discussed in the opening of this chapter.

MOVED BY READING

GLENBERG HAS DEVELOPED A reading comprehension intervention called Moved by Reading. In multiple studies, Glenberg has asked first and second graders to use toys to act out what is going on in a sentence while reading through the sentence itself. One sentence at a time, children read the text aloud, and then simulate its activity with physical toys. If, for example, the sentence is "The farmer drives the tractor to the barn," then the child places the toy farmer into the tractor and moves the tractor to the barn. As we've discussed, children learn best when they can purposefully create their experiences; it's the fastest way, per Piaget, for them to create reality in their mind. Moved by Reading makes the indexing of the text expression to what it means deliberate and direct. "Because the objects are physically present, they both prime the pronunciation of the words and help to constrain the objects to which the words can be indexed," Glenberg writes.[35] The child also needs to personally perform what's happening in the sentence, physically embodying the action of the sentence—who does what to what or whom—in his or her own actions. In this way, the child learns what the words mean and how sentences represent action. In experiments, these actively playing readers had over twice the success recalling the contents of sentences compared to the control group of children who looked at the toys but didn't play with them. In follow-up studies, Glenberg found that doing the same with cartoon toys represented on a computer screen yielded similar results to the kid playing with the toys firsthand. The intervention also has "the collateral benefit of helping children love to read," Glenberg says.

That second step of Moved by Reading is where the child is directed to simulate the action of a sentence in their minds. This is likely what's happening when we read anything: words orchestrate our own creation of dynamic imaginal worlds. The skilled writer, Glenberg argues, connects abstract symbols to embodied experiences. Consider first this long abstract description of centripetal force: The force acting on an object in circular motion is captured by $F = mv^2/r$, where m is mass, v stands for velocity, r is the circle's radius, and F is the sought-after centripetal force.

Now, here's Glenberg's imagistic example:[36]

> Imagine that you are on roller skates in a parking lot. To stop, you grab a post, and as you fly by, you start to spin around the post. That spinning is circular motion, and the force that you feel in your arms is centripetal force, that is, the force causing the circular motion. The speed of your skating before grabbing the post (v) will affect the centripetal force that you feel in your arms. If you are skating fast, then you will be jerked more vigorously when you grab the post than when you are skating slowly. That is the v^2 part of the equation: The faster you go, the greater the centripetal force once you grab the post (and the more it will hurt). Now imagine that you are wearing a heavy backpack (thus you have greater mass), but that you are skating just as fast as before. Will the force that you feel in your arms when you grab the post be greater or less than without the backpack? In fact, the m part of the equation indicates that the force will be greater: If you are more massive, then it is going to hurt more to grab the post than if you were not wearing any backpack. Finally, imagine that instead of grabbing the post with your hands that you have a rope with a loop, and you lasso the post with the loop while you hold on to the other end of the rope. If the rope is short, then you will be whipped around the post in a tight circle, whereas if the rope is long, your path around the post will be a more leisurely, large circle. In

which case will you feel more strain (centripetal force) on the rope and your arms? According to the equation, the radius of the circle (r) acts as a divisor so that the longer the rope, the less the force. You can get a feel for this by thinking about how much centripetal force you will feel while whipping around the post on a short rope compared to the more leisurely drift on a long rope.

This is an embodied account of how effective writing works. In every beat of that description, Glenberg is tying the abstract variable, which itself is meaningless to the student when first encountered, to an embodied experience that illustrates its meaning. It's really just a more sophisticated version of the farmer driving the tractor into the barn from Moved by Reading. Indeed, a strain of rhetoricians and scholars of the craft of writing have long been arguing for this imagistic clarity as a bellwether of accessible writing.

A good deal of the brain is involved in visual processing, and the most lucid writing caters to our visual nature. Vision is educated by how a baby interacts with things in the world. Relative to more opaque words, visual, imagistic words are easier to remember, both for healthy people[37] and also those with early dementia.[38] Indeed, the best writing—for adults included—is physical, concrete, and imagistic. As Steven Pinker—whose affinity for the computational theory of mind we eschew but whose writing advice we try to follow—details in his *Sense of Style*,[39] the hallmark of so-called classic style is to consistently orient the reader with concrete objects in the world. A lot of writing becomes impenetrable, whether in corporatese or academese, when the writer stays within the abstracted argot of their field, rather than concretely describing things happening in a manner that can be embodied or feel embodied. Hard-to-get-through writing is often abstract to the point of being experientially meaningless to the reader, while quality writing produces a constant play of images in the reader's mind. "Forcing yourself to describe things in concrete terms is a way of undoing your own idiosyncratic accumulation of abstractions," Pinker told us, "and to present things on the common ground that you're likely to share with readers. If I'm a psychologist and I say 'the infants were presented with a stimulus,' my fellow psychologists might know what that means, but if I say 'I showed Big Bird to a baby,' everyone knows what Big Bird means."[40] In effect, the most compelling prose is all effectively screenwriting: you have to give the reader scenes that play out in the mind.

Language allows us to share our thoughts and feelings, a transaction made possible by the fact that much of our *Umwelt* is shared by other people. The Austrian philosopher Ludwig Wittgenstein observed that "If a lion could talk, we could not understand him."[41] What he meant was that, without a shared perceptual world, meaningful communication is impossible. Wittgenstein felt that the bodies, ecologies, and concerns of lions and humans are too different for there to be a communicable sharing of experience between our species. In our own case, our *Umwelt* is not only shaped by our human body and natural ecology but also by our social environment. Our perceptual world is filled with concerns about other people and about what other people think about us. By virtue of our evolutionary history, to be human is to be social.

BELONGING

7.

CONNECTING

ACROSS THE UNITED STATES AND EUROPE, the volatile, violent first half of the 20th century was marked by a mysterious pediatric crisis. With the world wars and the Great Depression, the number of children in institutional care boomed. But those children were not doing well. Across ten asylums in the eastern United States in 1915, infants had a mortality rate between 31 and 75 percent by the end of their second year. A 1920 German report found that even in the best foundling homes, seven of ten infants died before turning one.[1] Following Louis Pasteur and the revolution brought about by germ theory, hygiene had greatly improved in these places. So had nutrition. Yet young children were dying at an alarming rate. Even in the finest institutions, infants had a 10 percent mortality rate. Affected babies had symptoms that looked a lot like depression, with infantile versions of dejection and disassociation. To an observer, these babies were wasting away.

The risk to life and health of prolonged hospital stays had been pointed out earlier and was described as a "failure to thrive" or *hospitalism*. In 1897, Floyd Crandall wrote in the *Archives of Pediatrics* "Even in general hospitals the attending staff soon learns that, except for certain incurable diseases, a prolonged stay is usually not advantageous. The younger the patient, the more marked does this become. Under one year, the death rate in all infants' hospitals is excessive."[2] It could be seen in any large urban children's hospital, he continued; it was more dangerous than pneumonia or diphtheria. "As this condition develops," Crandall noted, "progressive anaemia appears, and the child frequently dies from marasmus, or simple wasting without organic disease." Yet it was "not new"—fellow physicians had written about it years earlier. The state of institutional care for infants was even worse in the late 18th century: the Paris Foundling Hospital had an 85 percent infant mortality rate; its Dublin equivalent a 99 percent.[3]

"A study of hospitalism, and the strange tendency shown by infants to waste away in apparently congenial surroundings, naturally leads to an inquiry regarding its causes," Crandall reasoned, recommending "care, fare, and air"—as in enough, but not too much, physical touch; appropriate nourishment; and space to move in a well-ventilated room—as possible interventions. Yet Crandall was forced to admit that neither he, nor the medical profession, knew what to do about the failure to thrive of these young children. It was a brutal truth. Across the United States and Europe, excessive numbers of infants perished inside hospitals, orphanages, and foundling homes every year.

In 1947 an Austrian émigré psychologist by the name of René Spitz screened a short film to a group of psychologists and physicians at New York Academy of Medicine.[4] It was called *Grief: A Peril in Infancy*. Spitz showed footage, recorded with his own camera, of a range of small infants all with similar patterns of decline. "If the mother is removed from the baby for a few weeks or months, the loss for the baby is comparable to that of a school child suddenly orphaned of both parents, dumped on another continent, in an alien environment where nobody speaks its language and the customs and food are foreign," an introductory title card read. "It becomes increasingly unapproachable, weepy and screaming." In further months of maternal neglects, the infant becomes more psychologically withdrawn, and more physically rigid. In 37 percent of observed cases, Spitz reported, "the progressive deterioration of the total personality lead eventually to marasmus and death by the second year of life." Afterward, according to one recent history of the event, a prominent psychologist approached him with tears in his eye. "How could you do this to us?" he said.[5]

While Spitz would be criticized for methodological weakness and sloppy record keeping, the emotional force of his research—and, one might say, one of the first examples of documentary film pushing forward reform—would be tremendously impactful. His work, and that of the generation of researchers around him, would reveal that, for human infants, thriving is not simply a matter of getting enough calories, warmth, and other necessities of life, but also requires emotional and social nourishment. Caregivers are needed to provide warm affection and kindness. The physical health of children depended on it.

Humans are übersocial animals. Being socially connected is essential for health, well-being, and happiness. Paleoanthropology suggests that being immersed in a social environment with friends and loved ones is a default assumption of the human mind. We are born to belong.

Today, the need to maintain social connections is born out in public health data. Astoundingly, the data shows that having only a few friends or close acquaintances is more likely to make you ill and cause your death than such obvious health risks as smoking and obesity. Relatedly, socially isolated adults are more likely to smoke, less likely to exercise, and eat fewer fruits and vegetables. A large-scale survey of nearly 7,000 adults in the San Francisco Bay Area found that over a nine-year period, the risk of all causes of death—that is, the chance of dying for any reason—was more than twice as high for people with the fewest social ties as those with the most.[6] Similarly, people lacking social relationships, in the form of weak social support or in stressful marriages—are at greater risk for cancer, diabetes, and cardiovascular disease.[7]

Julianne Holt-Lunstad at Brigham Young University has recently brought the ill effects of loneliness into national focus.[8] In a groundbreaking 2010 study, she and her team assessed mortality data for more than 300,000 people followed for an average of 7.5 years. People with adequate social relationships—as in being integrated into a social network—had a 50 percent greater likelihood of survival during this period than people whose social relationships were lacking. On average, people with strong social relationships lived 3.7 years longer than less socially connected people. In fact, loneliness was found to be a greater mortality risk than smoking, excessive drinking, lack of exercise, obesity and air pollution. "The overall effect remained consistent across a number of factors, including age, sex, initial health status, follow-up period, and cause of death, suggesting that the association between social relationships and mortality may be general, and efforts to reduce risk should not be isolated to subgroups such as the elderly," she and her colleagues wrote. Several years later, her team followed up with an analysis of 3.4 million people worldwide, again finding similar mortality rates. Across the globe, and across demographics, isolation killed. "We wanted to know: Does it vary by country (It does not!)? Does it vary by cause of death (Doesn't matter!)? Is it stronger for men vs. women (Equally strong!)?" she said in an interview.[9] "This was a snapshot of real life, right? With implications for real-life health outcomes."

Just like the infants in the foundling homes of the 20th century, adults in the 21st century are in deep need of care. Holt-Lunstad compared the devastation of social isolation to the paradigm shift that came with the discovery of hospitalism, its mechanisms, and its treatment through bonding and affection. "To draw a parallel, many decades ago high mortality rates were observed among infants in custodial care (i.e., orphanages), even when controlling for pre-existing health conditions and medical treatment. Lack of human contact predicted mortality. The medical profession was stunned to learn that infants would die without social interaction. This single finding, so simplistic in hindsight, was responsible for changes in practice and policy that markedly decreased mortality rates in custodial care settings. Contemporary medicine could similarly benefit from acknowledging the data: Social relationships influence the health outcomes of adults."

While Western, and especially American, culture may venerate rugged individualism and self-expression, people in contemporary society depend on other people to survive, just as it's been through human history. The notion of "the individual" is a relatively modern invention, making its first stirrings in the writings of medieval Christian philosophers like St. Anselm and William of Ockham, and reaching its perhaps final form in free-thinking European Enlightenment.[10] The word *individualism* itself was not coined until 1815.[11] For the course of human history, the social group, the clan, the church, the state—these have been the fundamental elements of society. Our lives depend on other people, not just on the nuclear family but the larger communal group.

Throughout this book we have been exploring human ecology: how we fit into our environments and what they afford to us, the interplay of which is expressed in perception. Affordances are not only physical; they are also social. Other people offer possibilities for good (love, affection, support) and ill (threat, abuse, social anxiety). The brain reflects these social affordances. Through the course of evolution, the structures of social pain scaffolded off those for physical pain, providing one of the primary imperatives of our lives:

the drive to belong and be loved. To be embodied means not only being able to hold a stone but also being able to hold a hand. The most intimate of social relationships are built through touch.

THE CONNECTING TOUCH

THE LEGENDARY UNIVERSITY of Wisconsin psychologist Harry Harlow began a seminal 1958 paper on the importance of touch by commenting on the huge disconnect between the centrality of love in everyday life and its lack of study within research psychology. In the 1950s, psychologists assumed that animals, including us, were motivated by basic biological drives (like hunger, thirst, and pain) that motivated behaviors needed for survival (like eating, drinking, and withdrawing from sources of pain). So, where does love come from and the desire for social affiliation? Psychologists at the time surmised that we learn affection by becoming attracted to the caregiver who provides sustenance. Infants desire milk, caregivers provide milk, and thus over time infants learn to desire caregivers, or so the story went.

But this story did not sit well with what Harlow observed in his laboratory. In studies with infant macaque monkeys, he found that they grew attached to the cloth pads that lined their cages, hanging on to them and throwing tantrums when they were taken out to be cleaned. Conversely, a baby monkey raised on a barren wire-mesh floor "survives with difficulty, if at all" the first five days of their lives. This observation sparked the insight that contact comfort might be as important for infant health and well-being as basic biological needs.[12]

In what has become an iconic experiment, Harlow's team built two artificial mothers that stood in for the infant monkeys' real mothers. One mother was made of wood, covered by sponge rubber, and wrapped in terry cloth, with a light bulb radiating heat behind her. The other was made with wire mesh and a light bulb. One warm and fuzzy; the other warm and metallic.

For four monkeys, the cloth mother possessed a nipple that lactated, whereas for four others, the wire-mesh mother had the lactating nipple. The striking finding: in both groups, the infant spent 12 or more hours a day on the cloth mother, for the full 160 days of testing, with less than one hour spent on the wire mother—even if she was the source of food. This flew in the face of what had become received psychological wisdom: that nourishment drove behavior, with affection being an acquired outcome of feeding. Contrary to such views, it became abundantly clear that tactile comfort was deeply desired by infant monkeys regardless of whether it had been associated with food. "We were not surprised to discover that contact comfort was an important basic affectional or love variable," Harlow observed. "But we did not expect it to overshadow so completely the variable of nursing; indeed, the disparity is so great as to suggest that the primary function of nursing as an affectional variable is that of insuring frequent and intimate body contact of the infant with the mother." What is true for macaque monkeys must surely also apply to people, Harlow concluded, and hence he added: "Certainly, man cannot live by milk alone."[13] From the start, humans, like other primates, need to be touched.

THE SKIN IS A SOCIAL ORGAN[14]

LIKE OUR PRIMATE COUSINS, we express love and affection most intimately by touching each other. Nonhuman primates spend a considerable portion of their day touching each other through grooming. For example, baboons spend about 17 percent of their waking hours grooming other members of their group, even though the time needed for hygienic cleaning purposes would require only about 1 percent of their day.[15] The extra time spent grooming serves to reinforce social bonds and likely feels just wonderful to the recipient of this tangible attention. A baboon or other nonhuman primate will part the hair of another with a sweeping motion of one hand and pick out debris with the other hand. This sweeping motion is akin to a human caress. Mammals have specialized touch receptors in their skin tuned to this social touch.

Although our bodies appear not to be covered with hair, we are, nonetheless, as hairy as our evolutionary cousins. Human hair comes in two varieties: guard hair, which is thick enough to easily see, and vellus hair, which is fine, short, and hard to see from a distance. Men and women have equal amounts of hair on their faces. Male beards consist of guard hair. Women's faces are covered in vellus hair. Except for the palms of your hands, the soles of your feet, your lips, and your unmentionable body parts,* the rest of your body is covered with hair, mostly of the vellus variety. Wrapped around the base of your hair follicles are the caress receptors. They detect the hair movement caused by the sweeping motion of another's hand across your skin; and if the initiator of this touch is a loved one, then it can feel deeply pleasurable.

Caress receptors are the tortoises of our nervous system. They send their signals to the brain (via the spinal cord) very slowly. Now, because of their slowness, these receptors transmit the strongest signal when a caress is applied slowly with intermittent delays of a second or two. To precisely determine the optimal caress rate for these cells, a group of Swedish researchers placed electrodes on caress receptors in awake human volunteers. The researchers gently stroked the volunteers' skin above these receptors with a soft brush at different speeds. They found that stroking the skin slowly, at a speed of about ½ to 4 inches per second, was the ideal rate to evoke a maximal response from caress receptors. They then asked the volunteers to rate the pleasantness of brush strokes and varied the stroking speed. The most pleasing speed was found to be ½ to 4 inches per second, showing that the rate that causes the strongest response in the caress receptors coincides with the rate that feels most pleasant.

---

* Our genitals are hairless, although they are framed by tuffs of guard hair, which are a source of strong odors of sexual import.

We all intuitively know this. When caressing a loved one, you don't quickly rub your hand up and down their forearm nor do you rub it at too slow a pace. There is a Goldilocks zone—not too fast, not too slow—in which the speed of your caresses is most satisfying. We all discover this optimal speed as we share affection through touching and being touched by others.

Similarly to how pain tells you where something hurts while also motivating you to make it stop, touch receptors convey two sorts of information. First, they inform you about what has touched you (sharp, blunt, hard, soft) and where on your body the touch has occurred. Second, they convey an affective message (pleasant/unpleasant) having motivational content (keeping doing it/make it stop). The first messages are carried by fast-acting touch receptors and the latter by slow-acting cells like the caress receptors. These two receptor types follow parallel pathways to the brain, with the former terminating in the somatosensory cortex and the latter in the insula. The somatosensory cortex is responsible for discriminating and locating the touch. The insula is a brain area that is associated with affective feelings and motivation. It is this latter brain area that contributes to the wonderful feelings evoked by your loved one, Sweetie's, tender caresses. When caressed at just the right speed, volunteers in an fMRI brain imaging experiment show high levels of activity in their insula.[16] Researchers have also found that just watching someone being caressed at the Goldilocks speed—not too fast or slow—showed the same reward responses in the brain. So remember this the next time you're cuddling with someone you care about: a caress is an embodied message carrying deep emotional meaning.

HOLDING HANDS IN THE MRI

JIM COAN, NOW A colleague of Denny's at the University of Virginia, came to appreciate the vital importance of social support when he was doing his clinical psychology internship at the Southern Arizona VA hospital in Tucson. He was working with a World War II veteran who had late-onset PTSD. For most of his life, he had been symptom free. But now, as his condition grew worse, he refused standard treatment. His trauma was too much for him to even talk about. Coan tried relaxation therapy, where you ask the patient simply to breathe, but even just trying to relax was too stressful for his patient. Then he asked Coan a question, the answer of which that would forever change Coan's psychological perspective: "Okay, but can I please bring my wife to therapy with me?"[17]

Coan said sure, and she accompanied him the next time. Coan asked the patient to begin relaxation exercises, and again he resisted. Then his wife scooched her chair over and took her husband's hand. "It was like flipping on a light," Coan says. His patient started doing the exercises, and eventually he started talking about his horrifying wartime experiences. But he would talk only if his wife was holding his hand.

Coan was inspired to explore this emotional interdependence in the lab. In one highly cited experiment, Coan and his colleagues placed married, heterosexual women in a brain scanner. The researchers simulated a stressful situation by telling the participants that they were about to receive a mild electric shock in each of three conditions—while they were alone, while they were holding the hand of a stranger, and while they were holding their husband's hand. Holding someone's hand was associated with less threat-related activity in the brain, especially when holding a partner's hand. Most striking of all: the better someone felt about their marriage, the smaller was their fearful brain activation.

Several years later, Coan conducted a larger follow-up study with a more diverse set of relationships. For this hand-holding experiment, each of the 110 participants brought with them someone of another gender.[18] About a quarter of these pairs identified as friends, another quarter were people who were dating, another cohabitating, and another married. For the threat of shock trials, the participants either held the hand of their partner, whom they had brought with them; the hand of a stranger; or were tested alone. Once again, brain regions associated with threat became less active when participants held their spouses' hands, and more so for people who felt greater social support in their relationships. But there was a surprising result: this ameliorating effect wasn't just found with spouses but also with boyfriends or girlfriends, and with platonic friends as well. The strangers were another story: holding their hands actually led to a magnified threat response.

So what's happening when you hold the hand of someone you love? According to Coan, it's the "social regulation of risk and effort," which sounds abstract and theoretical but is actually vibrantly relatable. Basically, our companions help us spread the burden of life's many challenges. (Again, this is the sort of stuff that's reflected in idioms: *we're in this together, let me lend you a hand,* and the like.) Among other things, what Coan found was the special power of human touch—you'll feel more secure facing something scary when holding the hand of someone you love. All these findings rolled up into a theoretical framework that reconsiders some of the assumptions of how humans evolved through history and function today. It's called social baseline theory, and its principal claim is that the natural state of our species is to be embedded in a social context, surrounded by others who will share resources and contribute to effortful challenges.[19] Social baseline theory further asserts that our social nature alters "what the brain construes as the 'self,' expanding the self to include others in the social network or relationship"[20] What social baseline theory is saying is that we all expect to live within a social world—it's assumed and instinctual that other people will be around us; and whether or not they're there, and who they happen to be, shapes the personal world that we perceive.

FRIENDS LIGHTEN THE LOAD

SOCIAL BASELINE THEORY MAKES sense in Gibsonian terms: affordances change when you've got an ally to share a load, be it physical or emotional. This notion is borne out in studies investigating the perceived weight of boxes depending upon whether they are to be lifted alone or with another person.[21] On each trial, participants in these experiments were asked to estimate the weight of a box filled with potatoes. From one trial to the next, the weight of the box was altered by varying the number of potatoes that it held. After making their weight judgment, participants lifted the box either alone or with another person. It was found that prior to lifting the box, participants anticipated that it would weigh more if they expected to lift it alone compared to when they anticipated the help of another. Moreover, after lifting the box, participants judged its weight to be less if they had shared the effort with another rather than lifting it alone. In an especially clever experiment, the helping person was either healthy or obviously physically impaired. Participants judged the weight of the box to be greater when they anticipated lifting it with the impaired person compared to the healthy one. These findings show that having someone share the load doesn't just halve the burden, it actually reduces the burden's perceived weight.

In preceding chapters, we described Denny's research showing that hills look steeper to people encumbered with a heavy backpack compared to without the backpack. What about when the backpack-wearing participant is with a friend—would the presence of a socially supportive friend make hills appear less steep? This is the question that Simone Schnall* brought into Denny's office one day in 2007. At that time, Denny had not yet fallen under the influence of Jerry Clore; Jim Coan; and, of course, Simone Schnall.

---

* At the time, Simone was a postdoc at the University of Virginia. She is now a faculty member at the University of Cambridge, UK.

Denny was adamant that the presence of a friend would have no influence on the perceived steepness of hills. He was dead wrong.

The resulting publication had two experiments.[22] In the first, participants viewed a hill either with a friend or alone and it was found that hills looked less steep when a friend was present. In the second experiment, participants were asked to think about a friend, a neutral person, or someone they disliked. Again, it was found that thinking about a friend resulted in hills looking less steep relative to thinking about neutral or disliked persons; the latter two conditions yielded similar results. In addition—and this is reminiscent of Coan's hand-holding experiments—the better the quality of the social relationship in terms of relationship duration, closeness, and warmth, the less steep the hills appeared.

Friends lighten the load, both literally and metaphorically. In the case of perceiving hill steepness, it is not the case that a friend is going to push you up the hill. Rather, the presence of a friend means that, as opportunities and challenges present themselves, you can rely on your friends for help and support. Climbing the hill still requires that you expend resources, but the overall pool of resources available to you is expanded by including those of your friends.

No burden is more consuming and enduring than child-rearing. The expression, "It takes a village to raise a child," is a testament to the need for active community engagement in the care, rearing, and education of children. The "it takes a village" proverb may be of African or Native American origin; no one knows for sure. NPR quoted Lawrence Mbogoni, professor of Africana-World Studies at William Patterson University, as writing, "Proverb or not, 'It takes a whole village to raise a child' reflects a social reality some of us who grew up in rural areas of Africa can easily relate to. As a child, my conduct was a concern of everybody, not just my parents, especially if it involved misconduct. Any adult had the right to rebuke and discipline me and would make my mischief known to my parents who in turn would also mete their own 'punishment.' The concern of course was the moral well-being of the community."[23] Likely, the expression was conceived independently numerous times around the world as a universal prescription for the social support required to raise children. Humans do not—cannot—raise children by themselves. Our babies are too demanding for too long a time. Caregivers need help from others.

MOTHERS AND OTHERS[24]

SARAH BLAFFER HRDY, University of California, Davis professor emerita, has dispelled some of our culture's most deeply ingrained and male-centric assumptions about the family life of early humans. In the conventional view, likely born out of the Victorian assumptions of the era in which paleoanthropology first emerged, the human couple is the social unit of our ancestral line, with the man being a hunter and the woman being responsible for gathering edible plants, as well as handling all child care. In her research and writing, especially *Mothers and Others: The Evolutionary Origins of Mutual Understanding*, Hrdy has argued compellingly that this rugged-individualist vision of the human condition is woefully wrong and, in fact, would have been impossible to sustain for reasons having to do with evolutionary ecology and simple math.

Hunter-gatherer women bear children every three to four years. This is about twice the rate of other great apes. As any parent can tell you, human babies are extremely dependent, slow maturing, and expensive to raise. It has been estimated that it takes roughly 13 million calories.[25] Hence it was simply not possible for hunter-gatherer parents to feed and raise their children without help from others. Sharing the care of children with nonrelatives is called *alloparenting*, and while about half of other primates have some form of shared care, humans are the only great ape that do so—for orangutans, gorillas, or chimpanzees, only the mother will care for the baby. "For example, a mother orangutan will not allow any other individual to take her infant," Hrdy has said.[26] "She will be in constant skin-to-skin contact with that baby for at least the first six months of life, not a moment out of contact, and that baby's going to nurse for as long as seven years. This is a very single-minded, dedicated, exclusive kind of care."

Findings on contemporary human hunter gatherers drive home how important alloparenting is. Researchers studying the !Kung San people of southern Africa have found that when a baby cries, other people from the village come with the mother to attend to the child, and a third of the time, it's a trusted other that takes care of the baby.[27] A similar pattern has been observed among the Efé, a foraging pygmy group in the Democratic Republic of the Congo. The average infant will have about 14 different caretakers, with the mother in the primary role but the grandmother, father, aunt, brothers, and sisters all playing a role, as well as unrelated people in the village. Alloparenting is normalized from the start: the infant is passed around group members on the very first day of life, and at 3 weeks, the baby is with alloparents for 40 percent of the time, and at 18 weeks, the baby is with alloparents 60 percent of the time, spending the rest with the mother. Similar patterns have been observed in other foraging groups.

Hrdy posits that human babies, given the competition they have with their siblings, as well as the range of potential caregivers, have evolved to be precocious at determining who's going to give them care and how to elicit it. She argues that emotional attunement is something that evolution would have selected for. She spells it out in a thought experiment: "Start with an intelligent, bipedal primate with the cognitive and manipulative potentials and rudimentary theory of mind found in all great apes," she begins. "Rear this ape in a novel developmental context where maternal care is contingent on social support and where offspring survival depends on nurture elicited from multiple caretakers. This results in a novel ape phenotype in turn subjected to directional social selection such that over generations, those youngsters better at ingratiating themselves with others will be better cared and fed, and hence more likely to survive."[28] And who is not helplessly delighted by the sight of babies? In brain imaging research, activity in the medial orbitofrontal cortex, a brain region associated with reward and beauty, reliably occurs within a seventh of a second when looking at the face of an unknown infant but not an unknown adult.[29] Indeed, the more baby-like a baby looks—that is, rounder face, larger forehead—the more unrelated experimental participants are to find it cute and feel motivated to take care of it.[30] For babies, cuteness is a core adaptation to human ecology and its many affordances for being taken care of.

Hrdy's work ties the human ancestral past to the modern era. Child-rearing evolved as a multiperson endeavor, and the working mom has always been a part of the familial situation. Indeed, the isolated breadwinner and homemaker of the American mid-20th century may have been the leading model in only that era—a family arrangement that's a historical oddity, the extreme exception rather than the natural rule. Contemporary, large-scale demographic trends in developed countries attest to the high value of allomaternal support, and the necessity of combining working life and parenthood. Countries with state-subsidized alloparenting in the form of childcare, like France and the Nordic states, are replacing their populations, whereas those that don't support working mothers are at risk of, or are already experiencing, demographic declines—in Spain, Italy, and much of Mediterranean Europe, as well as South Korea and Japan, where the abandonment of schools in small towns has become a regular trend.[31] Indeed, the only reason the United States hasn't had a steep decline in births is because of immigrant populations, the same ones that are more likely to have allomaternal care.[32] When a society forces women to choose between becoming professionals or mothers, you get fewer of both.

THE POWER OF BONDS

IN THE 1940S, the British psychiatrist John Bowlby was trying to figure out what was so special about the bond between a baby and its mother. He was unimpressed by the psychoanalytic take, which involved a lot of "drives," a longing for the breast, and a focus on nutrition. Nor did he find much help in the behaviorist approach, which posited that beyond attending to basic biological needs, babies just needed sufficient stimulation. It wasn't the loss of emotional connection that made maternal deprivation so damaging, so the latter line of reasoning went, rather it was the lacked engaging experiences that caregivers provide.

Bowlby started his career during wartime England, and he found himself struck by the range of human life he saw at a school for maladjusted youths. Unlike what his Freudian peers insisted, Bowlby didn't think it was libidinous conflicts that were driving problems for these young people—it was the quality of relationships they had with their parents. Bowlby studied the juvenile thieves that populated the children's clinic to which he was assigned and found, almost without fail, that they had a troubled relationship with their mother and perhaps their father as well.

Anna Freud, the scion of her father's psychoanalytic tradition, referred to mother-love as "cupboard-love," since, according to the Freudian view, all the baby really needed was to eat. "I just didn't think it was true, I *knew* it wasn't true," Bowlby later observed. "There was a lot of fancy talk about breast-feeding and bottle-feeding and so on; I regarded it all as rubbish. It was completely contrary to my clinical experience. There were very loving mothers who had bottle-fed their babies and some very rejecting mothers whom I met in the clinic, women who were obviously very hostile, who had breast-fed their babies. And it seemed to me that the feeding variable was totally irrelevant, or almost totally irrelevant. So I was unimpressed with the conventional wisdom, but had nothing particular to put in its place."[33]

On a tip from a friend, Bowlby alighted upon the work of Konrad Lorenz, an Austrian ethnologist—that is, someone who studies animal behavior. Among other things, Lorenz was famous for his research on imprinting, or how some animals, like a duck or a goose, will latch on to the first thing they see when they emerge into the world. (In a classic experiment, Lorenz divided goose eggs into two groups, one group was left with the mother, and the other was placed in an incubator. The ones hatched with the mother imprinted on her; the ones that hatched in the incubator saw no mother but did see Lorenz. To confirm, he placed all the goslings under an overturned box, then let them out. Half ran to mom, the other half to Herr Professor.[34])

Ethologists had already established what they called "species-specific behaviors." These are a suite of behaviors that are common to nearly all members of a species, such as the barking of dogs, the meowing of cats, and the use language by people. Some species-specific behaviors have a critical period, meaning that, to be acquired normally, they need to be learned by an early age. A bird learns a song from its father, but only if that happens at the right time. A gosling imprints on its mother, but only within the right time window; otherwise the normal, functional bonding wouldn't take place.

Bowlby came to view the bonding between mother and infant as a time-critical species-specific behavior. The baby is compelled to coo, the parent to comfort. It's all about forming a mutual bond through a give and take of need and affection. Not having that bond, or having it form in some dysfunctional way, could hinder a child's development into a confident socially adept human being. "Bowlby introduced the formal term 'attachment' to describe the infant-mother bond," writes Robert Karen in his seminal history of the field. "Unlike bonding, which suggested an instantaneous event . . . attachment suggested a complex, developing process. Indeed, to Bowlby attachment was much closer to the idea of love, if not identical with it."[35]

We are born expecting to attach, and how attachment goes will frame the way we perceive the affordances of the social world of other people. The neuroanatomical nature of social pain speaks to this. Our social instincts are so engrained that their imperatives are enforced by the emperor of all emotions, pain.

Although Bowlby's views on attachment were focused on the mother-child bond, attachment has, in the intervening decades, broadened out into a range of anchoring relationships that develop over the life course. An "attachment figure" is someone who's a safe haven or a secure base—a person to whom we turn to when we are hurt, threatened, or anxious. It's theorized that as children grow into adolescence, they assert independence from their parents and move those attachment bonds to their friends. By the time adulthood is achieved, attachment roles have shifted from parents to peers, namely romantic partners.

Just as children discover how balls and rattles work by playing with them, they learn through their early social relationships how people work—and how they should act around them in order to get the love and nurturing they need. Bowlby, and his later collaborator Mary Ainsworth, would examine attachment and its styles empirically through their famous "strange situation" experiment. In front of a two-way mirror, a mother plays with her child, departs, and comes back, with observers noting the child's regard for the parent when she leaves and when they're reunited. The way that children cope with their mother's absence and reunion reflects whether their attachment to her is secure or not. While academics and clinicians vary on exact definitions, children exhibit several different attachment styles. Secure attachment has a grounded, stable quality. The insecure styles are anxious attachment, where the child seeks to be extra close; avoidant attachment, where the child is distant; and, as would be identified by Ainsworth and her colleagues in a later wave of research, disorganized attachment, where the child has unpredictable or frightened responses. To put it roughly, those styles are the product of how the child adapts to the mother's behavior. If a mother struggles with being autonomous (and allowing the baby to play on its own), then the baby is going to struggle with autonomy and want to stay close to mom. If the mother is dismissive of the baby's emotions, the baby will learn to quell its own.

It is estimated that about 50 to 60 percent of US adults are "securely attached," and are likely to be in a relationship and happy with it.[36] The rest carry around some kind of, often underlying, relational tension: some feel the need to be smotheringly close to their partner; others are easily spooked by vulnerability; and some people have such negative associations with intimacy that they don't want (or can't stay in) relationships at all. We also must note that these categories are not fate: "earned security" is what happens when someone who grew up socially insecure in some way is able, through a stable long-term relationship or perhaps sustained work with a therapist, to grow into a more secure version of themselves.[37]

Social baseline theory reflects our species-specific instinctive nature to perceive ourselves as embedded in a social world. Attachment colors those social worlds: you do not experience an objective reality of your romantic partners, friends, and family—you experience them through the lens of your own social upbringing. What kind of animals are we? Social animals. What do we perceive? A world filled with social opportunities and costs, weighted by one's attachment life history. The securely attached see more of the opportunities; the insecurely attached see more of the costs—biases that are self-fulfilling.

More contemporary research has sought out the evolutionary adaptiveness of anxious and avoidant attachment—which had, for at least a while, been seen as maladies to cure without value unto themselves. The Israeli psychologist Tsachi Ein-Dor has investigated this in ingenious experimental designs. In one, experiment participants were stationed in a room that slowly filled with nontoxic smoke—and the groups that included individuals with higher *attachment anxiety* escaped the danger more quickly.[38] In a related Ein-Dor experiment, participants were led to believe that a virus had infected their computers, and the groups with more *avoidant* people were faster to seek tech support. This provided evidence for Ein-Dor's hypothesis: both anxious and avoidant attachments have possible values to the group, alerting the rest of the clan to danger or being first to take action. Other recent research shows that children securely raised by more than one parent are better able to take perspectives of others, in line with Hrdy's writings.[39] Secure attachment triggers an "adaptive cascade," researchers say: if you're learning that it's okay to be distressed, then you're going to have an easier time handling the pressure that comes with taking a standardized test. Children that show signs of secure attachment at 9 months score better on standardized tests when they're 11 years old. We seem to transfer our attachment style onto our smartphones, too: socially anxious people feel naked without their phone; socially avoidant people want to keep the ringer off, thereby keeping communication behind a barrier.

COGNITIVE AGING AND SOCIAL NETWORKS

AS WE AGE, we slow down, both physically and mentally. This is hardly news, but it is surprising that cognitive decline is not just something that happens with old age. Strikingly, our cognitive abilities begin a slow decline in our 20s, the prime of life, and this decline accelerates as we approach and pass our 60s. But not everyone ages in the same way. People who exercise and are healthy preserve their cognitive faculties better than those who are sedentary and have health problems. In addition, there are life experiences that help individuals to be more resistant to cognitive decline. High among these is having a strong social network.

Cognitive decline is a natural part of aging. Like graying hair and wrinkled skin, a decline in our memory and reasoning abilities is an inevitable consequence of growing old. Dementia and Alzheimer's disease, on the other hand, are pathologies that occur with greater likelihood as we age but are the result of underlying diseases of the brain. And yet, even the cognitive impairments brought on by Alzheimer's disease can be lessened by having a strong network of friends and family.

One of the most comprehensive studies of Alzheimer's disease is the

National Institute of Aging funded Rush Memory and Aging Project, which recruits participants from the Chicago metropolitan area.[40] In a study on social support and cognitive functioning in Alzheimer's patients, researchers recruited 89 elderly people without any known dementia and kept tabs on them over the years until their deaths. These volunteers were tested every year with a battery of cognitive assessments. They were also interviewed about the depth and breadth of their social networks. Upon their deaths, researchers performed brain autopsies. As would be expected, poorer cognitive abilities late in life were associated with the magnitude of pathology revealed in the postmortem brain autopsy. But the size of the participants' social network was associated with less severe cognitive impairments. Participants who enjoyed large social networks showed high levels of cognitive functioning even though they exhibited large, pervasive brain pathologies.

It's a stunning, hopeful takeaway. People with the same level of brain pathology can have quite different levels of cognitive functioning. It depends upon how socially connected they are. Even the ravages of Alzheimer's can be kept somewhat at bay through the companionship of our friends and family.

Health, longevity, happiness, and cognitive functioning are all maintained and improved by social support. Holding a loved one's hand reduces anxiety. Hills look less daunting when ascended with a friend. The lesson of these findings is that the human *Umwelt*—our perceptual world—is a place filled with people who are perceived to care about our welfare and to be willing to aid us with our endeavors, expectations that were molded by our personal attachment experiences in child- and adulthood. As a species, we have evolved to be social animals. Caring for our babies is too difficult and time-consuming to be accomplished alone. As social animals, we see a world filled with opportunities and costs for social engagement. We identify as belonging to a specific group of people, our in-group, with the *out*-group on the *other* side.

8.

IDENTIFYING

ON AUGUST 11 AND 12, 2017, hundreds of white supremacists converged in Charlottesville, Virginia, to rally against the removal of statues honoring the Confederate generals Robert E. Lee and Thomas Jonathan "Stonewall" Jackson. Klansmen, neo-Nazis, and other far-right groups marched with torches, shouted racist slogans, and chanted, "You will not replace us. Jews will not replace us." They brandished semiautomatic rifles, Confederate battle flags, and swastikas. A 20-year-old white supremacist rammed his car into a group of counterprotesters, killing Heather Heyer, 32, of Charlottesville.

Like everyone he knew, Denny was stunned and heartbroken: How could such hatred and violence take place in his beloved community? For about a week, Denny assuaged his hurt by attributing the hate and violence to outsiders, people from other towns and other states, who had traveled to Charlottesville looking to provoke an incident. But he soon discovered that this self-serving explanation was misguided. A week later, at a dean's meeting of the College of Arts & Sciences' chairs and directors, the full depth and breadth of the situation was laid out for him by Deborah McDowell, director of the Carter G. Woodson Center for African-American and African Studies at the University of Virginia. She spoke of the racist conditions out of which the University of Virginia was built, has been sustained, and prospers up to the present day. The evil, she said, could not be so easily blamed on others; it has always been here and dwells among us.

The impetus for the white supremacist march on Charlottesville was a proposal by the city council to remove the statues of the Confederate generals. These statues were built more than 50 years after the end of the Civil War and their initial unveiling was accompanied by a Ku Klux Klan march through the city's downtown with a chorus of white townspeople singing "Onward Christian Soldiers." For an African American resident of Charlottesville today, what must these statues mean? Until the call for their removal, Denny had never asked himself this question. The University of Virginia was founded by an owner of enslaved people, Thomas Jefferson; it was built by the labor of enslaved people; and for years thereafter, students brought their enslaved servants to the university to serve them. In 2017, the grounds of the university provided nary a clue that these enslaved people had ever lived, died, and been buried here. This lack of recognition is now changing, hastened by the events of that summer. It took the words of Deborah McDowell to wake Denny up to the racial invectives, like the Lee and Jackson statues, that are hidden in plain sight throughout his community.

Bigotry is the flipside of altruism. It is an offshoot of the imperative "We take care of our own"[1]—there's only so much to go around. In-group versus out-group, my tribe versus others: these distinctions are the bases for not only cooperation and altruism but prejudice and violence too. Of necessity, humans are a social species. We can't feed and rear our children on our own, and thus we need social groups to assure their survival. For purely biological reasons, like the rate of reproduction and the time it takes for children to mature into self-sufficiency, it just can't be done. But if resources are scarce, then we might want to take care of our own first and not be so generous with others. So how do we define the out-group, the people whom we deem unworthy of our altruism and social support? We create contrasts between them and us. *They do it like this, but we do it like that.* The psychology literature refers to this pattern of opposites as *parochial altruism,* which proposes that aggressiveness toward the out-group and cooperativeness toward the in-group likely coevolved in people.[2] Adherence to a "take care of your own" creed often evokes the corollary proposition of not having to take care of others, which is based on the assumption that unqualified altruism would lead one into poverty or starvation. It follows from this assumption that if you do not want to give everything away, then you need to be choosy about whom to support. The need to protect limited resources causes people to adopt a hierarchy of social in-groups to which they belong and from whom they expect reciprocal altruism. These typically include groups like family, community, religion, country, and race.

This in-group versus out-group dynamic is a common currency of civic life. People will make sacrifices for the common good because of the webs of relationships we live within. It's one of the ways we define heroism: putting one's life on the line in the name of community, religion, or country. Often, these heroic acts are a response to some form of barbarians being at the gates—a despised out-group that is threatening the beloved in-group.

Intergroup conflict is one of the animating forces of history, and it's chilling to think about just how much human suffering it's created. Early evidence of intergroup violence goes back 10,000 years, to a group of two dozen people found to have been massacred near a lake in Kenya, and one imagines it extends back much further than what the fossil record tells us.[3] In the 20th century alone, more than 230 million people died in war, genocide, or some other form of intergroup conflict.[4] As the news relentlessly attests, intergroup conflict is also very contemporary. From 2002 to 2011, the world saw 104,000 acts of terrorism.[5] In 2018, the Southern Poverty Law Center counted more than 1,000 hate groups operating in the United States. Out-grouping is what allows for enslavement and genocide, whether in Germany, the United States, or Rwanda. It's also the animating force of nativism, whether Brexit, the white-supremacists rally in Charlottesville, or the resurgence of the far right that swung so many elections in the 2010s, from Brazil to Poland. But identifying whether someone is part of your in-group or belongs to an out-group is not simply a matter of what you think about them; it's also how you see them within your personal, social world. In-group/out-group identity is a matter of perception.*

HOW IN- AND OUT-GROUP MEMBERSHIP SHAPES THE WAY WE SEE

ON A CHILLY SATURDAY in November 1951, the Princeton Tigers hosted the Dartmouth Indians for a football game that would become the stuff of sporting—and psychological—history. It was the last game of the season for both teams. The Princeton Tigers had gone undefeated up to that point, led on the campaign by Dick Kazmaier, an All-American halfback who had just graced the cover of *Time* magazine, and would go on to win the Heisman Trophy, college football's highest individual honor.

---

* Throughout this chapter, we use the term *identity* to refer to social identity. Identity, of course, may be more broadly defined as related to one's personality, vocation, likes and dislikes, and so forth.

It was a brutal game from the moment of kickoff. Referee whistles blew and penalty flags peppered the playing field. Kazmaier, the star player, was forced to leave in the second quarter with a broken nose. In the third quarter, a Dartmouth player was taken away with a broken leg. Princeton won handily, 13 to 0. More impressively, the two teams combined for over 100 yards in penalties.

Controversy promptly erupted.[6] The respective student media of each school chastised the other team for dirty tactics in successive issues. The *Daily Princetonian* reported having never seen "quite such a disgusting exhibition." And "the blame," the paper continued, "laid primarily on Dartmouth's doorstep." In turn, the *Dartmouth* volleyed back, declaring that "the same brand of football condemned" by Princeton, that of roughing up the best players, "is practiced quite successfully by the Tigers." The game was a slugfest. But who started the fight?

These recriminations were (and remain) a classic spectator sport disagreement. To psychologists Albert Hastorf and Hadley Cantril—respectively of Dartmouth and Princeton—it was an opportunity to study conflicts in perception (and perception in conflict). A week after the game, the researchers administered a questionnaire to psychology students at both schools. While most of the Princeton students thought that Dartmouth started the brutal play, the majority of the Dartmouth students thought that both sides held a share of the blame. To avoid bias or misremembering, the researchers followed up by showing the actual game film to groups of students at each school, who recorded infractions and other impressions of the game as it unfolded before them. The Princeton viewers saw Dartmouth make twice as many infractions as their own team made. Meanwhile, Dartmouth students thought their team committed half the number of penalties that Princeton students thought they had made. And the Dartmouth fans thought that most of the penalties were a means for the referees to protect Kazmaier, the Heisman winner-to-be.

What to make of this? To Hastorf and Cantril, the bias to see more fouls committed by an adversary, as opposed to one's own team, reflected the fact that two people never experience the same exact objective "thing," but rather see things through the filter of their own desires and perspective. "The 'thing' simply is not the same for different people whether the 'thing' is a football game, a presidential candidate, Communism, or spinach," they stated in their conclusion.[7] The experience of an event, athletic or otherwise, can only be known by the experiencer; there is no perfectly objective perspective. This happens at a granular level. We pay attention to things that are significant to us, and if you're a Dartmouth fan, you'll be looking for more Princeton fouls and Dartmouth successes, and vice versa. In this way, your perceptions of an encounter, athletic or not, feels like a cohesive, even objective, account of reality. Hastorf and Cantril state this position with extreme elegance: "In brief, the data here indicate that there is no such 'thing' as a 'game' existing 'out there' in its own right which people merely 'observe.' The 'game' 'exists' for a person and is experienced by him only insofar as certain happenings have significances in terms of his purpose. Out of all the occurrences going on in the environment, a person selects those that have some significance for him from his own egocentric position in the total matrix."[8] Put in our terms: you do not see the world, but the world as seen by you. Where "you" are defined not only by the body that you have and the emotional state that you're in but also by the social group to which you belong. You pay attention to the things that are most relevant to you, and to the social groups to which you belong.

The Dartmouth-Princeton study marked the beginning of an ongoing and immensely illuminating strain of research into what's now called "motivated reasoning." The big idea deriving from this research is that perceiving and thinking occur within a mental world filled with biases, emotions, desires, beliefs, and other attendant concerns of the moment. Dan Kahan, a Yale law professor, notes that while we might assume that a sensory perception, like watching a football game, is independent of the irrelevant thoughts and feelings that spin around in our minds, it is not. Rather, these thoughts create motivations to see the world in particular ways. And crucially these are not "motives" in the traditional sense: "The normal connotation of 'motive' as a conscious goal or reason for acting is actually out of place here and can be a source of confusion," he says. "The students [viewing the Dartmouth/ Princeton football game] *wanted* to experience solidarity with their institutions, but they didn't treat that as a conscious *reason* for seeing what they saw. They had no idea (or so we are to believe; one needs a good experimental design to be sure this is so) that their perceptions were being bent in this way."[9] To put a twist on an idiom, it's not so much that *you'll believe it when you see it,* but *what you believe* shapes what you see.

Motivated reasoning helps to explain why people, especially political partisans, seem to live in different worlds and can look at the same thing and perceive completely different realities. In a recent experiment, Kahan and colleagues showed that people, quite simply, have a really hard time reasoning about facts that conflict with their political views.[10] Kahan's experimental participants were 1,111 US adults from diverse backgrounds. These participants were presented with problems requiring that they evaluate data from hypothetical treatment studies. (For example, participants might be asked whether a vaccine is an effective preventative for a disease based upon data reflecting the number of subjects who did or did not get the disease subdivided by whether or not they received the vaccination.) The participants' political orientations were assessed as were their mathematical reasoning abilities. It was found that math abilities predict performance on the test problems, an unsurprising result given that solving the problems required considerable mathematical skills. But this advantage for having high mathematical abilities was evident only for apolitical problems—like using data to compute the effectiveness of a skin cream for treating dry skin. When a similar type of problem was related to strong preexisting political beliefs—like using data to compute the effectiveness of a ban on concealed handguns for reducing crime—participants seemingly threw their mathematical reasoning abilities out the window when the data conflicted with their preexisting worldviews. If, for example, supporters of gun control were presented with data showing that a ban on handguns failed to reduce crime, then their performance dropped precipitously and their math skills were no longer predictive of their success in solving the problem. A similar stupefaction was found for the supporters of gun rights when the data confirmed the efficacy of banning handguns for reducing crime. Conversely, when the problems' solutions aligned with participants' political orientation, then overall performance shot back up and math skills were highly predictive of problem-solving success. The participants with weaker math skills were 25 percent more likely to get the answer right when it fit their political preferences; the people with stronger math skills were an astounding 45 percent more likely to get it right when it affirmed their views. "Being better at math made partisans *less likely* to solve the problem correctly when solving the problem correctly meant betraying their political instincts," observed the journalist Ezra Klein in his reporting of the study, in an article aptly titled, "How Politics Makes Us Stupid."[11] He continued: "People weren't reasoning to get the right answer; they were reasoning to get the answer that they wanted to be right." This finding flies in the face of what Klein characterizes as the "more information hypothesis": there's a perhaps especially American belief that if people just had enough data regarding some hot-button issue, then they'd see the light of reason and align themselves with the facts. If there were enough research and communication about an issue, the accreting evidence would be enough to sway opinion and policy. While admirable, thinking that all people need to change their minds is the right kind of information reveals a certain naivete in assuming that facts speak for themselves. What Kahan and his colleagues' work show is that we filter for and seek out the facts that are consistent with what we already believe. Kahan calls this is the identity-protective cognition thesis: we reason with ease about things that affirm our identity and avoid thinking about notions that conflict with the values of our social group. For example, if you grew up and remain in a community of fundamentalist faith, where professing otherwise would mean exile from the group, then you're likely to maintain those beliefs in order to maintain your good social standing. Seeing things from an adversaries' perspective is not easy, for, as Kahan's research shows, we become cognitively diminished when trying to reason with facts that conflict with our beliefs. Thinking—even doing math—is not a process happening in isolation but is instead embedded in our personal thoughts and group identities. There's a lot of meaning in the short phrase "identity-protective cognition"—namely that we use our reasoning and perception as a means of protecting our sense of self-worth, as well as the dignity of those organizations with which we ally ourselves, be they educational, religious, athletic, civic, or beyond.

In the introduction, we wrote the following about political polarization in the United States: "To some, a politician may be viewed as a champion of American values, whereas to others this same political figure may seem the worst sort of ignoramus. How can such divergent opinions develop among reasonable people living in the same world?" We suspect that many readers may have objected that people in this country are not living in the same world. People of different political persuasions get their news from different politically skewed news sources—MSNBC, Fox News—and consequently are getting different stories about what is going on in the world. This is true, but the research on motivated reasoning shows that even when presented with the same information, people interpret it differently. We literally become flummoxed, stupefied, and incapable of wielding our full faculties when confronted with facts that threaten our social identity.

THE OTHER "RACE" EFFECT AND DEINDIVIDUATION

UNLIKE POLITICAL AFFILIATION, NATIONALITY, or religion, race is assumed by many to be a genetically determined biological fact. Biologists, however, never even use the term—they don't subdivide groundhogs, for example, into different races. Biologists use the term *subspecies* to describe the genetic differentiation that may occur if different populations of a species become completely isolated from one another for a long period of time. This subspeciation has occurred four or five times for chimpanzees. Recall that chimpanzees rarely wander more than a couple of miles from their home nests. Given their stay-at-home nature, once different groups of these apes do become separated by some distance, they stay separated. Genetic mutations accrued independently in these isolated groups and eventually they've become sufficiently different such that biologists refer to them as subspecies.

Unlike chimpanzees, humans have always been and remain a species forever on the move. No group of nomads has ever walked so far away from other people that intermittent intermingling did not occur. Our wandering ways never caused our ancestors to completely lose touch with those they left behind. Because of this, there are no hard genetic divisions between people today that would justify referring to them as different subspecies or races.

In recent years, it has become easy to find out the approximate regions from where your ancestors hailed. All you need to do is send a small quantity of saliva to a commercial lab that will perform a genetic analysis and get an approximate sense of your ancestral origins. (This does come with some caveats, as the companies are comparing your genetic data against their reference libraries, which are not perfectly representative of the global population.[12]) Denny did this and found out that his recent ancestors came from England, Ireland, Germany, and Scandinavia. Does this imply that people from these different countries are of different races? Of course not. Small benign mutations that build up in a place will be slow to distribute around the world and are thus good indicators of where your recent ancestors resided. Genetic differences associated with recent place of ancestral origin are minuscule and lack any hard discrete genetic boundaries that would support racial definitions as far as biology is concerned.

People's appearances—especially skin color—are often taken as indicative of race. Skin color is another result of our wandering ways and the latitudes that our ancestors passed through. If generations live in sunny places near the equator like the tropics, then having dark skin is adaptive. Dark skin provides protection against ultraviolet light, which can destroy folic acid, and lead to birth defects. But if you migrate away from the equator, then dark skin becomes disadvantageous because you need some sunlight to produce vitamin D. The expression of white skin in "Europeans" is a relatively new phenomenon because access to the northern latitudes of Europe was blocked by glacial ice until about 13,000 years ago. The last ice age made the land uninhabitable for humans, and people did not wander into northern Europe until the ice had sufficiently receded about 10,000 years ago. These people had dark skin and were prone to getting rickets, a bone disease resulting from a lack of vitamin D, which is synthesized when the skin is exposed to sunlight. Selection pressures related to the production of vitamin D at latitudes that afforded limited sunlight brought about a prevalence of white skin in northern Europe. The genes for white skin, however, had already evolved over a million years ago in Africa in *Homo erectus*.[13] In fact, Lucy, our *Australopithecus afarensis* ancestor who lived over 3 million years ago, probably had white skin covered with hair, as chimpanzees do today. Shave a chimpanzee, and you will discover white skin beneath its fur. *Homo erectus* became dark skinned as they lost their fur and fell under selection pressures to protect their skin from too much UV light. The genes for skin color are older than our species and their selection for expression depended upon the recent migrations of our ancestors into extreme latitudes.

Skin color cannot be a basis for assigning racial categories. Instead, race is a social construct, is based upon people's beliefs, and has no grounding in human biology. As a social construct, race is more like the value of paper money or the veracity of religious claims: things that people accept as true, to varying degrees, but don't have a scientific grounding for. And as our findings on motivated reasoning have told us, if it's important to you that race is scientifically real for whatever reason, then you're probably going to think it's so, even after reading this.

Regardless of their source or truthfulness, however, beliefs are real to the person believing them; and, as we have argued throughout this book, they have real influences on our perceptions. For example, when encountering a new person, almost all of us use our racial beliefs to categorize them as black, white, Asian, and so forth, according to the ways we've been socialized. These racial categorizations come to mind unbidden and lead to, among other things, the *other race effect*, to which we now turn.

Stanford psychologist Jennifer Eberhardt traces her interest in racial bias to her primary school days, where she made what would end up being a crucial transfer between schools. She had lived in an all-black neighborhood in Cleveland, where all her significant relationships were with black people. Then her parents announced that the family was moving to a predominantly white suburb. When she got there, she made lots of white friends, but then realized something destabilizing. She was having trouble telling her white friends apart. How could this be?

"Our brains get attuned to what we're surrounded by," she explained in a recent interview.[14] "And so for me, I'm really good at recognizing black faces, being able to distinguish one from another. But then I moved to this other neighborhood where all of a sudden I'm surrounded by white people with whom I had never had any real meaningful interaction before. And even though I wanted to have friends and all of this in this new neighborhood I really couldn't tell their faces apart. I had been in really segregated spaces. I was attuned to different features, like [shades of] skin color. So it took a lot of practice in that environment before my brain was able to sort through [using hair and eye color]."

This was, Eberhardt would come to realize, a primary example of "the other race effect," wherein people from homogenous backgrounds have a tougher time remembering the faces of people from other ethnicities and seeing the nuances of their personal appearance, otherwise known as "individuating." (Not coincidentally, seeing people as individuals is a shortcut to empathizing with them.[15]) The problematically named other race effect is part of a larger category of out-grouping effects that shape the way we experience the world.

Most studies on the other race effect assess facial recognition. The research question of interest in these studies is: If you see someone on one occasion, then when you encounter that person again, will you recognize them as someone you've seen before? The basic finding from these studies is that you are far better at recognizing people who are in your in-group as opposed to your out-group.

As we first discussed in the introduction, these socialization effects start showing themselves remarkably early. At three months old, infants raised in a social environment predominantly of their own ethnicity prefer to look at people of their own group.[16] The other race effect is maybe the most extreme product of this socialization process. Researchers like David J. Kelly at the University of Sheffield who study this process propose a "perceptual narrowing hypothesis"—that infants attune to the faces of people most prevalent in their lives.[17] *Narrowing* is an operative term here: one experiment found that Caucasian three-month-olds equally recognized African, Caucasian, and Chinese faces, but by the time they reached nine months, they only recognized Caucasian faces.[18] But importantly, it is personal history, not ethnicity, that is the key factor. In another study, Korean children adopted by white European parents were found to have the same facial recognition patterns as a white French control group.[19] Following this pattern, research on multiethnic children finds that the other race effect is most likely to occur for groups to which the children have not been exposed.[20] In brain imaging tasks, people whose childhoods had more interactions with out-groups process out-group faces as more similar compared with people raised with little outside contact, as for example, Chinese Americans versus mainland Chinese.[21] All this research suggests that people are reporting on their experience when they say something like "You people all look the same." Their perceptions have narrowed by virtue of their life experiences, and consequently they are less able see people of other groups as individuals compared to those that they grew up with.[22]

Mix deindividuation with stereotypes and cultural assumptions about the indelibility of "race" and you get racial bias, the automatic sorting of people according to whatever beliefs you've received from the culture about "what they're like." Eberhardt has repeatedly shown how racial stereotypes influence perception, as for example, in the study in which white undergrads were primed by photos of black faces and subsequently more quickly recognized images of weapons and crime-related objects.[23] As mentioned in the introduction, white students in this study were shown either a rapid-fire clip of mixed-race images of men or one of just black men; and in the latter condition, they were quicker to pick out the silhouettes of crime objects like a gun or a knife emerging from a gray background resembling television static.

The key to defusing these biases is to get people to think less automatically. In a telling example, Eberhardt collaborated with Nextdoor, the free social network for neighborhoods and small communities that's in some 190,000 US neighborhoods. (Think Facebook if you were situated by your address.)[24] While the platform is principally used for neighborly things like finding a good plumber or a lost cat, the site also had a high incidence of "suspicious black man" postings. The startup wanted to address that racial profiling, so they talked to Eberhardt, who has been an adviser since February 2016. Following her advice, they modified their user interface by requiring users to be a little more specific about their suspicions. "For the crime and safety tab, you can't just write: There's a black man, suspicious. You have to identify some behavior that is actually suspicious," Eberhardt says. "And then be specific about what that person looks like so it doesn't sweep all black people in the same category." With the added friction of specificity, the platform saw a reported 75 percent dip in racial profiling. It has since added an online education portal for people to learn about bias and racial profiling and a "kindness reminder" feature that automatically asks users if they'd like to edit their postings if they're about to say something potentially offensive or hurtful to other users. Both were additions that Eberhardt had recommended.[25]

Fortunately, our poor memories for faces of other races appears to be somewhat plastic. People can be trained to better recognize other race faces. In a representative training study, white participants were shown images of very similar other race faces, either Hispanic or African American. Some of these faces were given unique names (e.g., Joe, Bob) and others were not. When names were provided, learning to identify these images required participants to pay close attention to individual facial features and as a result the other race effect was greatly reduced.[26] The other race effect also appears to be malleable depending on how you identify yourself. Black-white biracial people primed with their white identity more quickly identified white faces in a visual search task compared with when they had been primed to think about themselves as being black.[27]

But we need to continually remind ourselves that these studies on the other race effect are about the consequences of people's *beliefs* about race. Race is a purely social construct and is no more a category of the natural world than is the notion of nation state. And, just as national boundaries have changed over history—Texas used to be part of Mexico—racial boundaries have also changed: in early 20th-century America, people from southern Italy were considered to be black.[28] The boundary between black and white people is no more evident in nature than is the boundary between the United States and Canada. Yet, our beliefs about race and other group identities color all our social perceptions. We do not see the world, but the world as seen by us.

Otherization leads to embodied empathy blind spots. As was discussed in the Feeling chapter, brain imaging reveals that people, to a degree, really do experience another's pain as if it were their own. This tendency, however, has important individual differences. For example, people with high trait cognitive empathy—that is, folks who regularly imagine what it's like to be other people—are likely to respond strongly to videos of people being poked with needles.[29] In what's become a classic example, when watching one of these videos, it appears that the nervous system is simulating the sensory experience of having your own skin stuck through by a needle. But as a group of Italian researchers recently found, people don't have the same empathic response for those in an out-group. In that case, native-born whites living in Italy or African-born blacks in the same country were shown videos of an arm being poked with either a needle or a Q-tip. Someone of their same ethnicity prompted the empathic pain response, with the muscle group in their hand activating as if it were happening to them. When someone of a different ethnicity's arm was poked, this empathic response did not occur. And this in-group-specific pain embodiment effect was linked to scores on a common test for implicit racial bias,[30] suggesting that people felt the pain of the out-group more if they were less biased against them.[31]

But it's not that people felt an empathic reaction and then inhibited it—it's more an apathetic response to the pain of the out-group. Moreover, it's not just race that prompts these responses, or lack thereof. Fans of a local soccer team have been found to be similarly unempathetic when watching a fan of a rival team receive electric shocks.[32] It's not that you actively wish them harm, but rather you just don't notice when bad things happen to people of other groups or automatically get an embodied sense of what it's like to go through what they're going through. Like at the Dartmouth versus Princeton football game in the 1950s, the groups that we do and don't belong to shape what we notice, what we ignore, what we remember and forget.

BRINGING PEOPLE TOGETHER

WHEN PEOPLE ARE OTHERIZED, when they're considered or perceived to be part of an out-group, then they are psychologically opaque to us. They are, in a sense, objectified: treated as objects rather than as subjects.

Dwelling on the many ways that group identity pushes the way we see the world and other people can be a little deflating. But the ironic, and perhaps hopeful, thing about group identity is that it can change, and sometimes rapidly. Studies show that giving groups of children different colored T-shirts— blue for one group, red for the other—is enough to create in-group liking and out-group disliking.[33] Imagination can play a major role. Research shows that when people are asked to vividly imagine helping someone, they become more motivated to actually help other people.[34] Moreover, asking people to vividly imagine the inner experience of someone who belongs to a stigmatized group—putting yourself in their shoes—leads to greater empathy.[35]

Such laboratory finding give credence to the social power of novels like *Uncle Tom's Cabin* by Harriet Beecher Stowe, which helped energize the abolitionist movement, or *The Jungle* by Upton Sinclair, which threw light on the working conditions of immigrants in industrialized America. As the historian David S. Reynolds notes, Stowe wrote narratives specifically crafted to pierce traditionally southern assumptions about what was held to be the virtues of slavery—namely, that enslaved Africans were treated like family by the people who owned them.[36] The titular Tom is sold down the river, ripping him from his wife and children. He is eventually purchased by Simon Legree, a slave owner who is the novel's primary villain. Legree commands Tom to beat an enslaved woman, but he refuses.

Stowe, who grew up in a religious New England household, and whose brother became one of the most famous preachers of the 19th century, knew that her audience was Christian; and so she wrote in such a way as to evoke the experience of Tom as the noble sufferer, torn from his family, bearing beatings, but never renouncing his faith or inflicting cruelty on the people around him, even under the threat of his own death. The novel ends with Legree killing Tom, fulfilling the Christlike narrative.

The novel first ran in serialized form in the *National Era,* before being collected into two volumes. It was a blockbuster success, selling 10,000 copies in its first week, and then, over the course of its first year in print, sold 300,000 copies in the United States and a million in Great Britain.[37] It is hard to fully appreciate the sobering imaginal act that this put its many readers through: people who had perhaps abstractly thought about slavery as a bad thing now had a vivid account of the psychological experience of being enslaved, and one that magnified the Christian goodness of a member of what had been the ultimate out-group. It's a case study in what's described in Art Glenberg's Moved by Reading: describing the sensory experience of something, so that the reader is filled with an embodied, felt meaning, is the best way to drive home a truth. This is something Abraham Lincoln, whose fate was tied to Stowe's, understood implicitly: "He who molds public sentiment goes deeper than he who enacts statutes."[38] Perhaps that's why, when he apocryphally met her at the White House, he is widely reported to have said, "Is this the little woman who made this great war?"[39]

PEOPLE UNITE BEHIND A COMMON GOAL

IMAGINE BEING BORN INTO a port city in one of the world's great empires as that empire fell apart, then having it occupied by a foreign army not just once but twice, including razing three-quarters of the city's buildings, then combating the rise of nationalism in your own country, moving across the ocean for school, witnessing immense poverty, and then again being persecuted, this time as a foreign agent. And then publishing some of the most formative studies in the history of social psychology.

This was the life story of Muzafer Sherif, who was raised in a wealthy family in and around Izmir on the glittering Turkish coast. He was born into the collapse of the Ottoman Empire, saw the creation of the Turkish Republic, attended Harvard during the time of the Great Depression, and went to Germany to see a talk by Wolfgang Köhler, the noted gestalt psychologist, just in time to observe the rise of National Socialism. After finishing his doctorate Sherif returned to Turkey where he was an activist academic, opposing the discrimination against Jewish students and writing a book against racism. He was detained for four weeks in 1944 for connections to the Turkish Communist Party by the Nazi-sympathetic Turkish government, after which he left again for the United States, where he would live the rest of his life in effective exile. Seven years later, he would be investigated by the FBI.[40]

In a way that few people have been privileged—or fated—to have known, Sherif was privy to the many manifestations of social and intergroup conflict. But such conflict was not inevitable. Depending on the social conditions, groups could be brought together, and he would provide the evidence that demonstrated as much. So, in 1953, fresh off that brush with the FBI, Sherif received a massive grant from the Rockefeller Foundation, good for well over a quarter million in today's US dollars.[41] The plan was to assemble boys in a summer camp, set them against each other through any number of prods and prompts, and then finally bring them together again, thus illustrating the fluidity of group identity. The initial attempts to launch this massively ambitious study were disappointing failures. The boys quickly figured out that something funny was going on, with one boy reportedly asking the staff what the microphones hanging from the ceiling in the dining hall were for.[42] The key seemed to be that at the very start of camp, all the boys—aged around 11 or 12—were able to get to know each other, and *then* separated, causing much consternation among the children and a suspicion that the counselors—researchers posing as camp staff—were trying to manipulate them.

The third experiment worked, however, and became one for the textbooks. This happened in Robbers Cave, Oklahoma, with boys specially selected from across Oklahoma City by a researcher who sought out the most athletic, and thus most competitive, boys available.[43] This time, the children were kept separate for the first day. Then they were quickly divided into two rivalrous teams: the *Eagles* and the *Rattlers*. Following a loss in a tug-of-war, the Eagles burned the Rattlers' flag with matches that were likely given to them by a researcher. Then came vandalism, including camp counselors defacing a Rattler building to make it look like the Eagles did it and food fights that the adults in the room did little to stop. After much antipathy was stoked—lots of "sissy!" and "stinker!" hurled back and forth, and everything that could be competitive turned competitive, including pitching tents— Sherif and his team were able to bring the boys together, in the final phase of the study, under goals that members of both teams, regardless of their allegiance, could get behind. The adults had stacked stones on the water tank high on the mountain, blocking the water supply. This created an urgent problem that no single boy or small group could handle on their own, that of liberating the water supply, thus forcing the boys to band together to remove them. Similarly, the boys threw in their lots together to raise enough money to watch a movie, *Treasure Island*. These results fell in line with Sherif's realistic conflict theory: when there's competition over scarce resources, the fire of us-versus-them identities gets stoked. But when a shared obstacle or common enemy presents itself, warring factions ought to band together. Sherif called these shared objectives superordinate goals. This is a classic trope in science fiction: the aliens come to Earth, and humanity finally comes together. You can see this uniting under common cause as a reflex in civic life. It's well documented that after times of natural disaster, cities and regions will rally together, like the Cajun Navy that helped out with hurricane and flood relief along the Gulf Coast of the United States. (The Cajun Navy was a loosely affiliated, all-volunteer, mostly male set of boatsmen who decide to assist rescue efforts—though the practice has also been called vigilante disaster relief.[44]) Or consider that in late August 2001, George W. Bush had a percentage approval rating in the 50s. Following September 11, with the on-set of the War on Terror, it shot up to 90 percent—the highest for a president ever, greater than Truman's at the close of World War II.[45]

Yet there is risk here too: one of the most powerful superordinate goals is to frame a common enemy, which historically has often led to dehumaniza-tion. While you might think that "hate" is the operative word in the phrase "hate group," one should not underestimate the power of the group: if you want to prompt a sense of togetherness, single out a scapegoat—immigrants, the poor, minority religions or ethnicities, the people across the border or in our midst. You can't tell them apart; they're all alike. Your unity is ensured; their humanity is debased. Hence why global warming is such an effective way for humanity to destroy itself. There's no single group that you can blame for this global threat. The enemy is us.

OUR IDENTITY IS TIED to the people and groups to which we belong.

The sphere of our identity extends outward, through layers of social support, from the family, alloparents, neighbors, school, and religion to the extended homeland and culture. Every homeland is ultimately grounded in its unique geography, an ecological landscape that affords some kinds of sub-sistence living but not others. As will be discussed in the next chapter, it is the ways of life that geography imposed on our ancestors that form the bases for cultural differences that exist even today.

9.

ACCULTURATING

SOME THIRTY YEARS AGO, Richard Nisbett, then a rising star in psychology at the University of Michigan, became interested in murder and why it happens. A son of West Texas, Nisbett was curious about the American South and its long history of, and propensity toward, violent aggression. His own motivations were both professional and personal. He wanted to study culture using the tools of social psychology, and he wanted to start by studying the cultural environment in which he grew up. The prevalence of violence in the South was well-documented. As early as 1878, journalists were writing about how the South was more violent than the North. Today, well into the 21st century, you're three times more likely to be murdered in the Deep South as you are the North.[1]

Many explanations for the North/South disparity in violence have been offered over the years, like the legacy of slavery, the rate of poverty, or even higher year-round temperatures. Nisbett knew that slavery couldn't be the full story, because enslavement was most prevalent in the wet plains where cash crops like cotton grew well; and those regions, in fact, had lower homicide rates than the mountainous or desert regions of the South. To complicate the issue further, the gap didn't seem to be an urban issue. Cities in both regions were about the same in terms of violence. It was small southern towns that had higher murder rates than northern ones. In addition, black men were no more likely to be involved in murder in the South compared with the North. It was southern white men who had higher rates of homicide than did their counterparts up north. So what is going on with white southern men?[2]

A CULTURE OF HONOR

NISBETT AND HIS COLLEAGUE Dov Cohen at the University of Illinois dug into the FBI's homicide records. These logs provide demographic data about the perpetrator and victim, as well as crime-related context or cause, as for example, whether the killing was vehicular homicide, part of a break-in, or the fallout from an argument or affair. The research team sorted homicide incidents into two large categories: murders that involved some form of insult, like a love triangle or unsettled score, and those that didn't, such as murders associated with arson or robbery. They found that, compared with the rest of the country, murder rates for white men were higher in the South for insult-related killings but not for other murders. This was especially true for small towns where the incidence of insult-related murder was more than twice that found elsewhere: 4.7 per 100,000 people in the South versus 2.3 in the rest of the country.[3]

But the demographic stats weren't enough to make a conclusive case, so Nisbett and Cohen followed up on this statistical finding with an experiment, testing to see whether people from the two regions differed in how they responded to insults. It was a surreptitious study involving a staged confrontation. White male participants, originally from either the North or South, were asked to drop off a paper at a table, which was located at the opposite end of a long, narrow hallway. As each participant made their way down the hall, an actor unknown to them crossed the hallway in front of them and began sorting through a filing cabinet. The actor had to push the file drawer closed to let the participant walk by, and reopen it afterward. Then, seeing the participant walking back, he slammed the file drawer shut, walked toward the participant, bumped him on the shoulder, and audibly muttered "asshole" as he walked by. Two observers watched the whole exchange, and then rated how amused or angry the participant looked. As predicted, the participants from southern states were angrier after being bumped. Physiological measures corroborated this finding, with double the rise in cortisol and testosterone levels in insulted southerners than their northern counterparts. And these were southerners who had self-selected themselves to attend a northern school, the University of Michigan.[4]

These results, and others like them, support the conclusion that the South is one of several American cultures—and many worldwide—that abide by a *culture of honor*. Honor lies at the intersection of toughness, reputation, and masculinity, where any affront to one's honor will be met with swift and certain retribution. The governing principal: if you don't fight back, then you'll lose status among your peers. Various forms of honor culture are found all over the world and throughout world history, the drama of which have become touchstones of cinema and literature. It's shocking to see Tony Soprano maim someone for insulting his daughter. But actual historical examples can be even more brutal: in Sicily in the Middle Ages, the husband was *legally obligated* to kill his wife if she were found unfaithful.[5] The samurai of feudal

Japan lived by the comparably more stoic, but no less brutal, Bushido code: marked by resolve, loyalty, and death before dishonor, exemplified in the ritual suicide of seppuku, disemboweling one's self instead of surrendering.[6] Absent the presence of a sheriff, the old American West was so wild that it necessitated frontier justice, exemplified in Wyatt Earp's vendetta ride to kill the cowboys who had taken his brother's wife. The culture of honor is a familiar trope from Western movies: when you can't rely on the law, you have to take the law into your own hands. It's much of the same in areas of concentrated poverty in the United States: "the code of the street" is, in essence, a quest for respect.[7] Empirical studies of gangland murders in Chicago and Boston find that killings are motivated by retaliation, status seeking, and the collective memory of a gang, all of which are honor-related, symbolic behaviors.[8] It's the rule of retaliation: cross me, and you're gonna pay.

Where did the culture of honor come from in the first place? In a controversial explanation, Nisbett proposed that cultures were formed by the adaptations that our ancestors had to make in order to survive and flourish in the environments in which they lived. In essence, culture arose from geography.

A recurrent theme of this book is that people are prone to wander. If things are not going to your satisfaction where you are, then pick up; make use of your bipedal, endurance-animal body; and go somewhere else. And so members of our species walked out of Africa and eventually came to occupy the four corners of the earth. When they arrived at an appealing locale, they had to figure out how to sustain themselves. Following the first agricultural revolution, some 12,000 years ago, people had options available other than hunting and gathering, and these options depended on geography and climate. Some places afford planting crops, others herding domesticated animals, still others fishing, and so forth. If the ground is tillable, fertile, and rain is plentiful, then planting crops is a good choice; whereas if the ground is full of immovable rocks, then planting crops is going to be a losing proposition, and you're better off herding.

Most agricultural livelihoods benefit from people working together, but the presence of other people also raises the risk of certain types of misbehavior—like thievery. Given the agricultural opportunities afforded by a locale's geography and climate, social ways of life evolved to both maximize the benefits of social cooperation and also to defend against the malfeasance of others. These social norms, which arose as adaptations to particular places, created cultural worldviews—lenses through which the world and other people are viewed. Different landscapes support different kinds of livelihoods and these livelihoods set up cultural norms and rules for social relationships, all of which frame patterns of thinking and perceiving. These cultural worldviews last for generations, even after distant progeny have given up the ways of life that brought these worldviews into being. As the expression goes: You can take the boy out of the country, but you can't take the country out of the boy. Some cultures, like those that depend on rice, are socially interdependent—you have to work with your neighbor to get the next harvest. But others are socially independent—you can raise sheep more or less by yourself, but you need to keep an eye on your neighbors and make sure they don't rustle any of your livestock.

The historian David Hackett Fischer, in *Albion's Seed,*[9] traced US culture to four different groups of Britons that settled in various places in the country. Puritans went to the Northeast, the Cavaliers of southern England to Virginia, Quakers to the Middle Atlantic and Midwest, and the Scotch-Irish to the far West and the South. The Scotch-Irish are a key people for understanding honor. These people came to the United States and set out into what was then the frontier. In their original homeland, these Scotch-Irish immigrants had been herders; and herders are uniquely vulnerable to larceny, compared to farmers. "You're gonna have a culture of honor wherever your livelihood can be taken from you in an instant and the police are not around," Nisbett explained to us.[10] "That's the herding cultures of the world." The same thing isn't likely to happen if you're farming the land. It's much harder to make off with a field of grain in the night than it is to steal a goat. Honor culture, like any other culture, is the end product of social adaptations. Where people are vulnerable to theft but can't or won't call the sheriff, honor cultures take root.[11] The similar geographies and climates of Scotland and Ireland can support herding but are much too rocky to afford extensive crop planting. So the Scots and Irish took to herding and, of necessity, became vigilant about defending their animals from possible theft. The honor culture that this vigilance embodies is an adaptive consequence of having to make a living in a rocky place. The progeny of these Scotch-Irish herders settled the American South, they brought their culture of honor with them, and this worldview persists among certain populations to the present day.

CULTURAL RELATIVISM

IT MAY SEEM UNINTUITIVE and hard to accept that the way you view the world is influenced by the agricultural practices of generations of people who came before you. We fail to notice the impact of culture on the way we think for the same reason that we fail to notice our own accents when we speak. Everyone speaks with an accent, but we only notice accents in other people. Even when visiting another country, it always seems to be the other person who is speaking oddly, never us. Similarly, we don't experience our cultural worldview because, like accents, we only notice differences in how others behave, implicitly taking our own worldview as being the way it is.

But remember, you do not see the world as it is but rather the world as seen by you. This mantra, which we've repeated throughout the book, is a critique of naive realism, that pesky assumption that our experience of the world is exactly the same as everyone else's. So if we are going to have a better understanding of ourselves and our fellow human beings, we need to appreciate the startling individuality of everyone's experience. Nisbett, and his collaborators around the world, scale up this claim to encompass cultural diversity. At times, psychology seems blind to cultural relativity, by assuming that research participants, drawn from a pool of university undergraduates, are universal stand-ins for the rest of humanity.

The field of psychology grew out of the European intellectual tradition, and its assumptions about the laws of human nature may say a great deal about European habits of mind. Nisbett encountered his own cultural biases, in much the way you can walk face-first into a glass door—you didn't realize the barrier was there. In 1980, Nisbett coauthored a book with what he now effacingly refers to as a "modest" title, *Human Inference*. Roy D'Andrade, a cognitive anthropologist, read it and told the author that it was "good ethnography," by which he meant that Nisbett had nicely articulated the worldviews of North Americans and Europeans while completely ignoring those of other cultures. Nisbett relayed the anecdote in a review paper two decades later: "The author was shocked and dismayed. But we now wholeheartedly agree with D'Andrade's contention about the limits of research conducted in a single culture. Psychologists who choose not to do cross-cultural psychology may have chosen to be ethnographers instead."[12]

Since its beginning a century and a half ago, psychology has concerned itself with universals, aspects of human nature that are common to all human beings. Psychology's preoccupation with finding universals was inherited from its parent, philosophy. John Locke, David Hume, and John Stuart Mill all wrote about cognitive processes in a manner that assumed them to be the same for all people. Twentieth-century psychology also adopted universality, thereby leading researchers to look for unbending rules of mind and brain that applied to people regardless of context. Universality was embraced by the computational metaphor of the mind that was ushered in with the cognitive revolution starting in the 1960s. "Brain equals hardware, inferential rules and data processing procedures equal the universal software, and output equals belief and behavior," Nisbett and his colleagues observed.[13] Sure, the beliefs and behavior could be different, given the different inputs that individuals and groups have, but the underlying structures—the "source code," to extend the metaphor—was the same. These "basic" processes, like categorization, learning, and reasoning, are assumed to be identical among groups. Whether you were in New York or Shanghai, the essential functions of the computer-like brain are assumed to be the same. You might use different languages, but what lies under the hood is interchangeable.[14]

But psychology's cousin, anthropology, had spent decades, if not centuries, documenting the staggering diversity of human customs and ways of life, especially in the form of ethnography, where the fundamental assumptions of a group are made explicit through long-form interviews and immersive observation. As the two fields began to cross-pollinate, Nisbett and others would use the tools of social psychology to study culture experimentally. While he'd spent the early part of his career assuming that cognitive processes were universal across people, Nisbett was intrigued by the anthropologists and philosophers who claimed differently. Then, crucially, he was given opportunities to work with Chinese and East Asian students on cross-cultural research projects. The resulting findings would cause Nisbett to reformulate his ideas about how the mind works and spur a revolution in the field of cultural psychology.

ANALYTIC VERSUS HOLISTIC THINKING

A CLASSIC INSTRUMENT IN cultural psychology is the triad task, in which participants are presented with three objects and asked to select the two that are most similar to each other. Imagine, for example, that you are shown a picture of a cocker spaniel, a Labrador retriever, and a head of cabbage. You will likely group the two dogs together because they are, after all, both dogs. In a seminal 1972 study, fourth- and fifth-graders from Indiana and Taiwan were presented with trios of several different sorts—kinds of people, furniture, tools, food, vehicles, and the like. They were asked to group the two that went together and then give a reason for why they made their choice. When presented with a picture of a man, a woman, and a child, American kids put the man and woman together because "they are both adults," while the Taiwanese children grouped the woman and child together because "the mother takes care of the baby." Similarly, when presented with a chicken, a cow, and grass, the Americans put the chicken and cow together (both animals), while the Taiwanese put the cow and the grass together (cow eats grass). Hence a principle was revealed: the American kids sort things by categories, whereas the East Asians sort by relationship.[15]

In other studies, participants—mostly college students—were presented with a trio of illustrations like those of a hand, a glove, and a scarf. Which two form a pair? Is it the hand and the scarf? The hand and the glove? The glove and scarf? Most Westerners will group the glove and the scarf together, since they're both varieties of winter clothing. They belong to the same category—a sort of reasoning that's the hallmark of *analytic* thinking. Most Easterners will match the hand to the glove. The glove protects the hand; the hand fills the glove. They have a relationship. (As we will see in a bit, the East and West distinction is too sweeping, but initially this is how these findings were interpreted.)

The triad task is beautiful in its simplicity and remarkable in the way that it uses a simple pairing exercise to sample the workings and structure of a much more complex yet coherent cognitive process. Whether you're matching the man with the woman and the glove with the scarf or the child with the woman and the hand with the glove depends on what's been variously called a *cognitive* or *thinking style* and whether the culture you're from (or immersed in) has an analytic or holistic worldview. Analytic and holistic reasoning represent deep assumptions about the philosophical and metaphysical nature of reality, and they also govern our most basic day-to-day behavior, like how we interact with strangers. These are foundational assumptions about how the world works.

In the holistic view of the world, things are constantly blending into one another; in an analytic view, it's either this or that. The law of the excluded middle, a cornerstone of Western philosophy, asserts that any proposition is either true or the opposite is true. For example, the proposition, "Denny has red hair," is either true or its opposite, "Denny does not have red hair," is true—you can't have truth both ways. (In fact, Denny does not have red hair.) For Aristotle, claiming that a statement cannot contradict itself and still be true is an absolute necessity for knowing anything.[16] In Eastern thought, however, self-contradictory statements can express truths. For example, consider this statement from the ancient Indian philosopher Nagarjuna, one of the founders of Mahāyāna Buddhism: "Everything is real and not real, both real and not real, neither real nor not real. This is the Lord Buddha's teaching."[17] To analytically minded readers, such statements may seem like gibberish.

As is probably apparent, given the names that we're bringing up, analytic and holistic thinking have their seeds in two ancient cultures that had little to do with each other, the *East* and the *West*. These two cultures are thought to support two quite different worldviews: Western individualism— the rugged individual—versus Eastern collectivism—the nail that sticks out gets hammered down.

The historic cultures that gave us these radically different ways of seeing the world emerged in vastly different geographies that supported vastly different ways of life. Greece is mostly landscapes plunging into the sea. Your ways of life are herding, the market, and seafaring. If things didn't go well for you in Athens, then you could wander off to another city-state or maybe hop on a ship to the islands. To a degree unique among ancient civilizations, the Greek had a sense of individual agency, at least if *he* was a free man with land. He showed individual prowess in war and on the verbal battlefield of the marketplace and debate floor—places of individualistic confrontation that lay at the center of Greek civilization. Ancient China was the Middle Kingdom. Outside of that was desert (Gobi), mountain (Himalayas), and seas (East China and South China). Where the Greek landscape afforded individual freedoms, the Chinese landscape commanded collective continuity. While there were periods of debate, at least among scholars, the emphasis in Chinese culture was (and largely still is) on harmony. Your "individuality" was seen in terms of being part of a larger social system and its many obligations. You were in your village and family for the long haul, and these social and familial relationships conferred both benefits and obligations. You lived an interdependent life, and so your mind was conditioned to pick up on the many lineaments of relationship dynamics.[18]

The affordances of different geographies led to differences in how people construe the self: independent or interdependent. Western cultures are perhaps unusual in that they view the self as "bounded, unitary, stable, and separate from the social context," as one recent consortium of international scholars have noted,[19] where other parts of the world see the interdependent self as "closely connected to others, fluid, and contextually embedded." That leads to vastly different cultural outcomes: people with independent selves would seek to express, actualize, and differentiate themselves from others, while the interdependent would want to fit into relationships and maintain harmony. One worldview promotes actions that benefit the individual and their personal goals, whereas the other encourages people to work toward the common good.[20]

The philosophical writings of the East and West reflect the collective wisdom of their culture. The pronouncements of sages from both cultures provide a record of the societies in which they found themselves. Quite tellingly, Confucius elaborated on the mechanizations of social obligations, whereas Aristotle was concerned with universal truths. What's so amazing is that the cultural styles of thought associated with these philosophers are still so much with us today, even in the very way we pay attention to things.

Nisbett and his collaborator Takahiko Masuda asked Japanese and European American college students to watch realistic animated scenes of underwater life, and then describe what they had seen.[21] The first thing that Americans reported was usually about the fish, whereas the Japanese described the setting. For example, an American might report, "The trout was swimming to the right"; whereas a Japanese person might say, "There was a lake or a pond." Remarkably, both groups spoke about the fish with the same level of detail. However, the Japanese students made twice as many observations about the fish's relations with inanimate things in the scene ("The big fish swam past the gray seaweed") and they made a whopping 70 percent more statements about various details in the background of the environment. The holistic person looked at the whole scene; the analytic person attended more to focal objects. Consequently, Japanese students had a tougher time recognizing familiar objects in new settings, which didn't affect American students as much.

To Nisbett, the social imperatives of a holistic culture affect how you attend to the world. "If you're in an interdependent culture, you're looking to the horizons, the corners of the room," he told us in a follow-up interview.[22] "Other people are salient to you in every moment in a way that they aren't in an individualist society. If you're looking for other people, incidentally, you pick up what's on the horizon and the corner of the room. Your perceptual habits end up with you seeing context, where the individualist sees the particular individual or machine he's trying to deal with. If you're scanning the horizon constantly for people and their interactions, you're picking up lots of other things as well. If you're casting a wide net perceptually, you pick up relationships that someone in an independent culture wouldn't pick up on. The more objects in the environment you're attending to, the more relationships you're going to see."

These differences in what people of Eastern and Western cultures are inclined to notice—either the whole scene or focal objects, respectively—are evident in the histories of these cultures. Surgery has long been a part of Western medical history, since it naturally follows from analytic thinking that if you could just find the part of the body that "isn't working," then you could "go in and fix it." Chinese medicine, instead, took a more holistic approach in which invasive surgery was seen as repulsive. Early Western physicists thought that the properties of objects caused their motions. A stone falls to the earth because of the gravitational attraction between the mass of the stone and that of the earth. A block of wood floats in water because the wood's material is less dense than water. The Chinese appreciation of interrelationships, on the other hand, led to early discoveries of magnetism and acoustic resonance.[23]

But dividing the world's cultures into two categories, East and West, is far too simplistic, and today, finer distinctions are made.[24] For example, research finds that, like East Asians, Brazilians are inclined toward holistic thinking but with greater optimism about the future and more varied emotional expressivity—suggesting that Brazilians, and perhaps Latin Americans more generally, are as holistic as East Asians but without a Confucian heritage.[25] Russians and Malaysians have been found to be more holistic in their habits of categorization, visual attention, and the like than Americans or Germans.[26] There are within-culture differences as well. For example, Japanese on Hokkaido, the rugged northernmost island of Japan, are more independent than their main-island peers.[27] Northern Italians categorize things according to taxonomy more often than do the more holistic, southern Italians.[28] Turkish villagers who farm are more interdependent than those who herd.[29] Within the United States, working-class adults show more holistic thinking than do those in the middle-class.[30] In so many ways, the culture we inhabit shapes our perceptual world.

CHINA: RICE VERSUS WHEAT CULTURE

WHAT DO YOU DO when you run into a problem? Do you actively attempt to take control of the situation, or do you adjust your behavior to accommodate to what's going on, gently avoiding possible conflict? While these behaviors are, in part, a matter of individual differences—agreeable folks are more likely to accommodate, extroverts are more likely to assert themselves—the behavior is also driven, researchers have found, by culture. In the classic view, individualists, at the far end of social independence, will bend the situation to fit their will, and collectivists, who are more socially interdependent, will alter themselves to fit the situation. In studies, Americans are more likely to try to control the situation, whereas Japanese are more inclined to adjust to it.[31] This difference is suggestive of another East/West divide, but recent research suggests otherwise.

After living in China for six years, Thomas Talhelm came to realize that its people did not uniformly conform to two popular narratives about East Asians. First, not all Chinese people were inclined to behave in an accommodating way; and second, not all urban Chinese were more individualistic than were their rural counterparts. Now a faculty member at the University of Chicago's Booth School of Business, Talhelm had lived in China while a student and freelance journalist. During his time there, he noticed something odd about two cities in particular: Guangzhou in the south and Beijing in the north. Both popular media and academic research suggested that there existed a cultural divide between urban and rural Chinese worldviews. By this account, Guangzhou, formerly known as Canton, and Beijing, the Chinese capital—both booming, hypermodern East Asian cities—should be culturally similar. If you followed the prevailing theory that urbanization inclined people to adopt a more individualist worldview, then the people in both Guangzhou and Beijing should share this cultural perspective. However, in

Talhelm's firsthand experience, having spent years living in cities across the Middle Kingdom, they did not.

When Talhelm was in Guangzhou, he would go grocery shopping and found that the aisles were always narrow and the stores reliably crowded— it's a city of 12 million people, after all. He'd inevitably bump into his fellow shoppers, and when that happened, they would often tense up, cast their gaze to the floor, and walk away deferentially. It felt like people were avoidant of confrontation. Such accommodating behavior was not so prevalent when he ventured north. "Literally the first day I arrived in Beijing, I took a taxi from the airport into the city," he told us. "At the destination, the taxi driver pulled over to let me out. As he did, he stopped in the bike lane. An old man biking by was not happy about it. As I got my big bags out of the back, the old man stood and yelled at the driver. He yelled so long I remember him pausing, as if he had to think about what else he could yell. Meanwhile, I felt partly responsible for making the driver take more abuse because I was taking so long to get all my bags. That was quite the introduction to the north."[32] The "avoid confrontation" way of life, which was so much a part of life in the south, was clearly less important up north.

Within China, people from Beijing are known to be chatty. It's common to talk politics with a cabdriver, which is unheard of elsewhere. Beijingers are more welcoming of domestic transplants, and generally thought to be tough, made stout by hard winters.[33] In the south, so the stereotypical assumptions go, people are shyer around strangers and try to avoid conflict. Talhelm realized that these north/south differences were at odds with prevailing theories of Eastern culture and the effects of modernization. A critical insight came to Talhelm in 2008 while he was taking classes as a freelance journalist. The teacher showed a map of differences in dialects in modern China. In some places, the character *shou* (手) means "hand," and in others, it means "arm." He remembers being shocked by what he saw in the map: "Rather than being random or blob shaped, the difference split almost perfectly along the Yangtze River," he recalls. "North of the river it was 'hand.' South of the river it could also mean 'arm.' I thought, 'I bet that's the outline of these north-south differences I've been experiencing!'" Because if people speak the same, then their cultures are probably quite alike. "I didn't know what that river represented at the time, but I suspected it was key to the differences I was seeing."[34] The Yangtze River, originating in Tibet, wends nearly 4,000 miles through the cultural middle of China, bisecting the country into north and south. To the north of the Yangtze, wheat-cultivation dominates agriculture, whereas rice is king in the south. Difference in how these two crops are cultivated, Talhelm realized, could be the difference maker that he was looking for.

Growing and harvesting rice takes a lot of work, more than a single family can manage without help. The communal work required is a bit like barn raising, which used to be common in rural America. You cannot hoist the frame of a barn by yourself, so you ask your neighbors to help out, with the understanding that you and your family will gladly reciprocate later on. Rice farming is like that, but all the time—it requires twice as much work as farming wheat. In fact, records from medieval China attest to the prohibitive workload of cultivating rice, with one farming guide from the 1600s advising, "If one is short of labor power, it is best to grow wheat."[35] Rice also requires standing water, which demands irrigation, an innovation unto itself, and one that's interdependent by nature. There is only so much water to go around, so how you use your water affects your neighbor's usage. The planting terrains need to be built, dredged, and drained, an affair that requires lots of people working together. To handle these demands, villagers from China to Malaysia have formed cooperative labor exchanges, where planting and harvesting are timed so that one neighbor can help the next. In such a highly interdependent social endeavor, cooperation is imperative—a norm to be enforced. Wheat cultivation relies on rain, so there is no need for dredging, and not nearly the same need for neighborly help, save for the occasional barn raising. Compared to the rice farmer, the wheat farmer can be reasonably self-reliant, although not quite to the degree as the sheepherder.

Agricultural practices also influence culturally defined gender roles. For example, whether sowing seeds requires a plow affects how men and women engage in the cultivation of grains. A plow can be used only on deep, even, and nonrocky soil, and is best suited to crops that require lots of land to be prepared over a short period of time, like wheat, barley, rye, wet rice, and teff. A heavy and cumbersome plow demands considerable physical strength, and plowing is typically a man's job. Crops like maize, sorghum, roots, or fruit trees can be cultivated on less amenable soil, require less land, and can be planted over a longer period. These crops also require less physical strength to sow, such that either sex can do the work. Historically, the plow-heavy cultures have stricter gender roles, and the plow-light ones enjoy greater equality. These differences in agricultural practices, which came into being long ago, continue to affect gender roles today. For example, in the plow culture of Pakistan, 16 percent of women work, and in Burundi, a non-plow culture, 90 percent of women work.[36]

That notion that cultural worldviews grow out of the subsistence farming practices of a region is the core argument for what Talhelm calls "rice theory." In administering tests to more than 1,100 participants across China—from Yunnan to Xi'an to Beijing to Hunan—Talhelm's research team found strong evidence for this theory. First up was the triad test, the same scarf-glove-hand matching task that we discussed earlier. As predicted, the southern Chinese, drawing from their rice culture heritage, made more relational groupings—hand to glove. Wheat-culture northerners were more inclined to analytical pairings—scarf and glove. Then came a sociogram test, where participants are asked to draw a diagram of themselves and a few of their friends. Unbeknownst to the participants, the researchers were interested in the size of the drawing representing the self versus those drawn of the friends—with the relative size differences serving as a measure of individualism or self-inflation. (Tellingly, Americans draw themselves about 6 millimeters larger than their peers, Europeans 3.5 millimeters larger, and Japanese slightly smaller.) Among Talhelm's participants, northern Chinese drew themselves 1.5 millimeters larger, not unlike Europeans, while southern Chinese drew themselves smaller, not unlike Japanese.

To see whether these experimental results obtained with college students were reflected in general societal patterns, Talhelm dug into relevant population-level data, such as the number of patents filed and divorce rates among northern and southern regions, from 1996 to 2010. The north had more patents (a sign of individualist innovation) in 1996, but that leveled off into the 2000s—likely thanks to the tech boom in Guangzhou and elsewhere. Incredibly, the divorce rate was 50 percent higher in the north in 1996, and that difference held as the overall rate rose into 2010. "It is a safe bet that none of our thousand participants have actually farmed rice or wheat for a living," Talhelm writes. "Instead, the theory is that cultures that farm rice and wheat over thousands of years pass on rice or wheat cultures, even after most people put down their plows."[37] Talhelm's study not only made it into *Science* in 2014, one of the most prestigious scientific journals, but also served as the cover story. The headline read "Cultivated Minds: How Rice Farming Has Shaped Psychology."

In an exceedingly clever follow-up study, Talhelm simply pushed a chair into people's way in northern and southern Chinese coffee shops. An unknowing customer would walk into the coffee shop, and because there was no demarked path, they would have to pick their way through café tables to get to the counter. Talhelm positioned an unoccupied chair so that it blocked the most direct path from the entrance to the counter, which was a rather rude thing to do. The customer's choice was to either find a circuitous path around the tables to avoid the chair or take the direct path to the counter by pushing the chair out of their way. Talhelm found that in Beijing, assertive northerners tended to push the chair aside, whereas in the southern city of

Guangzhou, people were more inclined to accommodate to the situation and find a path around the offending chair.[38]

In yet-to-be-published work, Talhelm has found that students who move away for college—from a wheat culture to a rice culture—begin thinking more holistically by the end of their first semester, even if they've moved to a megapolis like Shanghai. He's found similar results in India, where rice is grown in the east and south, and wheat in the north and west. The message is clear: these crops, and the social constraints that they place on cultivation, have tremendous downstream effects. Humanity's great cultural divisions are not between Orient and Occident or East and West. Rather, cultural worldviews arise from how the early inhabitants of a region pursued their subsistence living: Did they need to cooperate with others or could they mostly go it alone?

SOCIAL AFFORDANCES AND RELATIONAL MOBILITY

COLLECTIVISM, AS IT'S TRADITIONALLY understood, has certain warm and fuzzy connotations—being loyal, nurturing relationships, looking out for the people around you—which stand in stark contrast to the me-first, lone-wolf qualities of individualism. Yet in day-to-day interactions—interpersonal warmth, especially toward strangers—the tone is the opposite. In his travels around the globe, Drake discovered the stereotype that Americans, despite coming from a place of supposed rugged individualism, are notorious for telling their life stories within twenty minutes of meeting you; whereas, if you're going to make friends living in Seoul, then it'll likely be with fellow expats, or international-leaning Koreans. It's an odd irony: individualists are forthcoming, collectivists aloof. What gives? As scores of economists, psychologists, anthropologists, and the like have noted, social relationships differ across cultures. They can be radically dissimilar within a country, state, or even the same city. In some places it's easy to make friends, change jobs, find romantic partners. In others, you're basically enmeshed from birth in the same web of long-term relationships.

The ease with which people can form new social relationships is called, "relational mobility." To date, most of the work on this cultural dimension has focused on forming friendships. Friendships exist in societies across the globe, but their nature varies with locale. In low mobility places, your friends are a matter of circumstance—the people with whom you grew up or went to school. In a highly mobile environment, friendship is much more of an active choice, a kind of mutual contract between people who share the same interests. If you're in a low mobility place, friendships are more guaranteed and stable, but it's also harder to depart from them and acquire new friends (or a new partner or even a new job). This tension is reflected in language: in Germany, one has many *Bekannte* (basically "acquaintance" but ever so slightly warmer) and a handful of long-term *Freunde* ("friend," but you only get a few). Mobility also helps explain why cultures can be so incredibly different in terms of how easy it is to meet people. If you're somewhere where it's super easy to make new friends, it makes sense to bare your vulnerabilities a little bit faster, since the mutual exchange of confidences makes people feel close and forms lasting bonds. However, if your social resources are already spoken for by the crew with whom you grew up, then you're not looking to make new friends.[39]

Like much of cultural psychology research, most studies of relational mobility have looked at only a handful of countries, often contrasting the Far East and North America. To address this, a team of researchers—27 of whom share authorship, including Talhelm—set out to conduct a truly global study. Between 2014 and 2016, they ran advertisements on Facebook news feeds in 39 different countries looking for people willing to take a quiz about friendship or romance. For example, in English, the sponsored post was under the banner "World Relationships Study" and read "Find out how your best friendship stacks up. 5 minute quiz with instant feedback. Take the friendship quiz!"[40] The countries they chose were determined by engagement on the social media platform, to ensure the researchers found enough participants, as well as representing as many geographic and cultural variants as possible. The results were, for the first time, a global index of how relationships work across the planet. North America was highly mobile, with Mexico more so than the United States or Canada. South America was found to be quite mobile as well—a fascinating result, given the holistic reasoning style previously found in Brazil, which is traditionally associated with more interdependent, low-mobility cultures. The sampled countries within the Islamic world—North Africa, the Middle East—were all relatively low in relational mobility. East Asia was also low in relational mobility, with Japan the lowest of all. The Anglophone countries of Australia and New Zealand were relatively high. Perhaps most surprising of all, Europe proved to be quite diverse: Germany, Estonia, and Turkey were all somewhat below the global mobility average, with Hungary more so. France and Sweden were highly mobile, and Spain, the UK, Poland, and Ukraine were above average. Portugal sat at the average.

Where did this great continuum of human culture come from? The research team found that having a history of threats, be it in the form of inclement weather, disease, population density, or poverty, was linked with lower social mobility, with Taiwan, Morocco, and the Philippines on the far end. The low-threat countries were the most socially mobile, like Mexico, Canada, and Sweden. The other factor was interdependent/independent agricultural subsistence style—with Japan, Taiwan, and Hong Kong being on the extreme end of interdependence and least mobile, and with Mexico and Brazil being the most independent and mobile. We see again a cascade of causes: history, geography, and climate select for certain ways of life, and these lifestyles give rise to certain kinds of social interrelationships. The adaptations, which were long ago required to survive and flourish in a given ecology, shape the way we perceive our social worlds in the present day.

EPILOGUE

THE PATH IS MADE BY WALKING

WHERE HAVE WE BEEN?

ABOUT 375 MILLION YEARS AGO, a fish called *Tiktaalik*—an Inuit word meaning "large freshwater fish"[1]—raised its head out of shallow tidewaters, pushing itself up by its fins. Over millennia, these fins evolved into limbs as this fish, or another like it, initiated the cascading evolution of vertebrate land animals that includes reptiles, amphibians, dinosaurs (including avian dinosaurs, birds), and mammals. Fins became limbs and life on land was off and running (and crawling, galloping, and even flying). The brains of these animals played catch-up, evolving under new selection pressures that arose from new ways of acting in new environments. The body led the way.

Fast-forward to a little under two million years ago, when one of the branches of the vertebrate tree of life walked upright on two legs across the African continent and eventually wandered into Europe and Asia. This was our immediate ancestor, *Homo erectus*—upright man. These people looked an awful lot like us. They were about our height and had our long legs, upright posture, and dexterous hands. They possessed a body very similar to our own; however, their brains were only about 60 percent the size of ours. Our ancestors' bodies provided opportunities for action that a bigger brain would eventually realize. As we've seen throughout this book, our brains evolved to take advantage of the opportunities that our bodies afforded in the circumstances in which we found ourselves. And as we've also said before, we did not first evolve a big brain and then hands that could realize the brain's desires; rather, we first evolved hands and then the neural machinery needed to exploit their potential. Our brains evolved in service of our bodies. There is a reason why *Homo sapiens*—wise man—was preceded by *Homo erectus*—upright man. Had we not taken to bipedalism, we likely would never have become the clever, brainy animal that we are today.

As discussed in chapter 2, bipedalism freed the hands to become dexterous instruments of manipulation. *Homo erectus* possessed these hands, but these people made few artifacts with them. They hunted large animals, their sole weapon being a pointed stick. Only much later did it occur to anyone to affix a sharp, pointed flint to its end. *Homo erectus* was a persistence hunter, chasing bigger faster prey for hours in the hot, midday sun until the animal collapsed from exhaustion and could be easily dispatched with a pointed stick. Although these people were slow in comparison to their prey, they possessed a magnificent cooling system—sweat glands almost everywhere on their bodies and a notable absence of insulating fur—that their furry prey did not have. Bipedalism conferred another endurance advantage. Doggedly pursuing their prey, persistence hunters were one of the most annoying animal on the planet, *Homo vexas*. The extensive sweat system evolved by *Homo erectus* had another fortuitous consequence: it provided the cooling system needed to keep our yet-to-evolve big brains from overheating. Brains are voracious energy users, and before big brains could evolve, an adequate cooling system needed to be already in place.

Our species evolved from *Homo erectus*—perhaps via an intermediary step, *Homo heidelbergensis*—about 200,000 years ago. Eventually, these people started to act like us as they learned how to modify their environment; to make artifacts of increasing complexity and utility; and to create new perceptual experiences through artificial means, such as music, language, and art. Like *Homo erectus,* our species wandered out of Africa, and over tens of thousands of years, we came to occupy every habitable niche on the planet.

WHERE ARE WE GOING?

IT IS IN OUR nature to pick up and leave if we are not happy with where we are. Our species evolved during times of climate change in Africa, and those people who survived to become our ancestors walked away from their homes when the water supply dried up. Today, our technologies make it possible to leave our physical environment by becoming immersed in artificial ones.

People enjoy using artifice to create novel perceptual experiences that could not be realized in other ways. In rough historical order, examples include music, language, art, architecture, and writing. These were followed by media, or artificial sights and sounds that present content like news and entertainment. More recently, interactive social media has become pervasive, where everyone's an author of content. Current innovations include virtual reality (VR) and augmented reality (AR), where the real or artificial body is immersed in artificial worlds. We can think of no reason why it should not be possible to create any sort of perceptual experience imaginable with a fidelity indistinguishable from natural experience. Not with today's technologies, not real soon, but someday.

Today's VR headsets* are based upon a design that was first built at the NASA Ames Research Center in the mid-1980s,[2] and as luck would have it, Denny's NASA collaborator, Mary Kaiser, shared an office with the VR lab director, Scott Fisher. Denny fell in love with VR, not only because it is so much fun but also because it affords research opportunities to create controlled visual environments that can be actively explored, unlike the passively viewed computer displays that were used by most visual scientists at the time.[3] In collaboration with computer scientist Randy Pausch, Denny built his own VR lab at Virginia in the mid-1990s. Also through his collaboration with Pausch, Denny got to participate in Walt Disney Imagineering's creation of a VR *Aladdin* ride, and he flew a magic carpet through a virtual re-creation of the *Aladdin* movie.[4] Such a perceptual experience would have been unattainable fifty years ago. What sorts of experiences will artists and engineers make possible next?

Not only can VR place you within any imaginable virtual world, it can also give you any imaginable virtual body, called an avatar. Now given that you have gotten this far in our book, you will not be surprised to learn that when your body is changed, so are your perceptions. In a paper entitled, "Being Barbie," Swedish researchers used VR to put people—including Denny—into the body of a Barbie doll, which caused them to perceive their surroundings, and the objects therein, as proportionally larger than when their avatar was the same size as their own body.[5] During a visit to the Karolinska Institute lab in Stockholm, Denny donned a VR headset, and when looking down he saw that he had become a Barbie doll. To heighten the illusion, one of the researchers, Henrik Ehrsson, repeatedly stroked Denny's leg with a small soft paintbrush, and simultaneously, Denny saw his Barbie doll avatar's leg being

---

* Immersive VR systems consist of a pair of small digital display screens set within goggles worn over the eyes, called a head-mounted display or HMD. The HMD presents stereo images of computer-generated virtual worlds. The position and orientation of the HMD are tracked so that the virtual scene remains fixed in space as the person in the virtual environment moves about.

stroked by a paintbrush. The touch sensations from his own leg coincided perfectly with the sight of the brush gently touching the avatar's leg. After a couple of minutes of this, Denny acquired a compelling sense that the doll's leg, as well as the rest of her body, was his own. Surveying his surroundings from his newly acquired Barbie perspective, objects appeared larger and farther away than when viewed with a normal-size avatar. Denny felt like he had become a Lilliputian in Gulliver's world, revealing again that the body is the measure of all things.

Nick Yee and Jeremy Bailenson, of Stanford University,* have aptly named the perceptual changes that are evoked by becoming embodied in virtual avatars, the Proteus effect, after the Greek god Proteus, who could change his shape at will.[6] In one of Yee and Bailenson's studies, participants were given either attractive or ugly avatars; and while immersed in a virtual world, they answered questions from an experimenter of the opposite sex who was located behind a curtain in both the real and virtual environment. Relative to those with the ugly avatars, participants having attractive virtual bodies stood closer to the experimenter and provided more intimate details when asked to talk a little bit about themselves. In another experiment, they varied the height of avatars and found that participants with taller virtual bodies behaved more aggressively in a negotiation game than did those with shorter avatars. This tendency—for people with taller avatars to engage in more aggressive social interactions—transferred to a real face-to-face social encounter that occurred after participants had completed a VR experience that manipulated avatar height.[7]

The Proteus effect is showing promising utility in medical, social, and legal domains. Numerous VR approaches are currently under investigation for pain management.[8] For example, some patients with swollen limbs experience pain relief when they view their avatar's affected limb shrink in size.

* Yee was a Stanford PhD student at the time of this research. He is now at Quantic Foundry.

Racial bias is subject to the Proteus effect: implicit racial biases are reduced when light-skinned people embody a dark-skinned avatar in a VR interactive world.[9] Moreover, this reduction in bias persists when participants are tested a week later.[10] In another study, it was also found that, after being embodied in a dark-skinned avatar, white participants were more cautious in determining the guilt of a black defendant in a mock courtroom trial.[11]

Other investigations of the Proteus effect exploit VR's potential to put you into any imaginable body, including those of famous people. Mel Slater, of the University of Barcelona, and his colleagues immersed people in the virtual body of either a white-haired Albert Einstein or an anonymous person. Those who were "being Einstein" performed better on a cognitive task and showed reduced implicit bias against older people.[12] In another study, Slater and colleagues created a virtual scenario in which people could switch back and forth between being a mental-health therapist or a patient, and thereby give counseling to themselves about a personal problem. In one condition, the therapist's avatar looked like themselves, and in the other it looked like Sigmund Freud. When "being Freud" the moods of participants improved.[13]

What sorts of perceptual experiences do we want to have? Into what sorts of artificial environments do we want to wander and with what bodies? These are deep and consequential questions that reflect on human nature. Effective restorative health environments, for example, invariably entail interacting with nature. This need to be in nature seems to be as fundamental as is our need for social affiliation. A few years ago, Denny was asked to help select and create artwork installations for the new Emily Couric Clinical Cancer Center at the University of Virginia. The committee's goal was to provide artwork that would reduce stress and promote healing. There is substantial research literature suggesting that people respond best to nature.[14] You can't beat a walk in the woods or a stroll down a beach. With support from local philanthropy, Denny installed a superlarge, flat-screen TV in a waiting room. Slowly changing images of the Blue Ridge Mountains were presented with 3D surround sounds of gentle winds, rustling leaves, birds, and such. The goal was to provide a virtual window onto an inviting natural scene into which one's mind could wander.

WHAT WILL WE DO WHEN WE GET THERE?

THERE ARE MANY WELL-RECOGNIZED risks associated with the rise of smart technologies—think robot takeovers in *The Terminator* or *Matrix* movies. What is less obvious about their increasing prevalence is how they will affect our own abilities and goals.

The rise of laborsaving machines during the last century and its contemporaneous impact on bodily weight gain serves as a cautionary example. The rise in the incidence of obesity is due, in large part, to a decrease in caloric expenditure. In general, offloading human labor to machines did not result in people finding other forms of exercise. Rather, people have become more sedentary, which has contributed to an obesity epidemic. If sitting is the new smoking, then what will be the impact of cognitive laborsaving machines on our cognitive skills and endeavors?

The decline in people's ability to perform unaided calculations—to do simple arithmetic—should serve as a warning. Many undergraduate students in Denny's classes cannot perform simple arithmetic calculations, of the nature of 18×7, in their heads. They require a calculator. (The same goes for Drake, his millennial coauthor.) This is so even though a calculus course is a prerequisite for declaring a psychology major. Similarly, defaulting to GPS to get you from point A to point B is a recipe for not remembering how you got there. As a consultant for tech companies, Denny has recommended the use of landmark-based navigation instructions—"turn left after you pass the school on the right, go up over the hill"—that encourage the user to attend to the surrounding landscape and thus more easily commit the area to memory. If we want to learn and remember, we have to be cognitively engaged.

It has been argued that artificial intelligence will free us to take on those tasks in which people excel over machines. Thinking long-term, however, there may not be any tasks that, in principle, machines can never do better than people. We don't mind so much when machines take over our physically demanding chores. Forklifts are good. On the other hand, we were not so pleased when a chess-playing computer program beat Garry Kasparov, the world's chess champion, in 1997, or when a Google artificial intelligence program AlphaGo toppled the human champion of the ancient East Asian game Go in 2016. How will we feel should music-composing programs surpass human composers, or—we shudder to think—a computer program writes a better book about human psychology than the one you are now reading?

Yet people will always strive for independence. A desire for self-efficacy is evident in a common plea of the very young: "I want to do it myself!" Children balk at adults' efforts to do too many things for them. They reach a point when they want to tie their own shoelaces no matter how slow and awkward the process. Similarly, the ill and infirm balk at having others take too much control over their lives. Providing patients with opportunities to be independent, no matter how small and seemingly insignificant, benefits their health.[15] Agency is a requisite for health, intellectual growth, and self-fulfillment.

A lesson of this book is that doing—willfully acting with your body—precedes knowing. The kitten that drives the carousel learns to perceive the affordances of space, but the passive passenger kitten does not.[16] In a future world in which AI is ubiquitous, what will we choose to do for ourselves, and consequently what will we know as a result?

Antonio Machado, a Spanish poet writing in the early 20th century, wrote:

Traveler, there is no path.
The path is made by walking.[17]

The path is our footprints. The body leads the way.

ACKNOWLEDGMENTS

DENNY

Debbie Roach, my wife, is a professor of biology at the University of Virginia. Most of what I know about evolution and its relevance for psychology I learned from her. Graduate students are the creative engines of research. Those who contributed to the ideas presented in this book include Jonathan Bakdash, Mukul Bhalla, Sarah Creem-Regehr, E. Blair Gross, Sally Linkenauger, Cedar Riener, Jeanine Stefanucci, Lars Strother, Elyssa Twedt, Rebecca Weast, Veronica Weser, Jessi Witt, and Jonathan Zadra. Our lab's research contributions are as much of their making as they are of my doing. I have benefited and learned much from wise and caring colleagues and professional friends: Tom Banton, Bennett Bertenthal, Jerry Clore, James Coan, Beth Crawford, James Cutting, Hunter Downs, Traci Downs, William Epstein, Arthur Glenberg, Mary Kaiser, Michael Kubovy, Randy Pausch, and Simone Schnall. For roughly 30 years, I've taught a course, Introduction to Perception, at the University of Virginia. Over the years, students in this course asked questions that I could not satisfactorily answer and that forced me to broaden my perspective and question some of the underlying assumptions of my field. This book is the result.

DRAKE

On the first page of *Maps of the Imagination: The Writer as Cartographer,* the University of Houston creative writing professor Peter Turchi says that the process of writing occurs over two interrelated phases. First the writer is explorer; second, the writer is guide. I had first started exploring the topic of embodiment in 2014, gathering string for a possible book by interviewing primary researchers in the field. One of those interviewees was Dennis Proffitt. Little did I know that we would partner up for what would become a years-long endeavor in exploring the field and its many neighbors, finally arriving at the present tour of what it means to be embodied and evolved as we humans are. I am grateful for this adventure, and I hope we have been adequate guides to the realm of perception.

Our book argues that humans are communal creatures, and I have relied on my professional, writerly, and personal communities to complete this project. Thank you to the many friends and colleagues who read various versions of the proposal and book chapters, including Adam Grant, Emily Esfahani Smith, Stephen Sherrill, and Jack Cheng. I'd like to particularly highlight the help of John McDermott, who read a startling amount of rough copy and offered generous edits, and Jess Jackson, who introduced us to our agent.

More personally, I am blessed to have many persons in my life who have supported me in the long exploration that this book has taken. My guides include my sister, Liza; my brother, Grant; my mother, Nicole; my creative compatriots Ryan, Kevin, Gustavo, Ellie, Becca, and many others; and most crucially my partner, Gabriela, who I am quite sure is excited to have me free to do things on weekend afternoons now that the book is done.

ACKNOWLEDGMENTS

DENNY AND DRAKE

Carol Mann and her assistant, Agnes Carlowicz, guided the writing of the book's proposal, and Carol found it the right home at St. Martin's Press. Through rounds of revisions, our editor, Daniela Rapp, directed our rewriting; and through her skill and patience, the book became more focused, economical, and fluid.

FURTHER READING

*If you'd like to dive deeper into how your body
shapes your experience, we recommend these titles.*

James J. Gibson, *The Ecological Approach to Visual Perception* (Boston: Houghton Mifflin, 1979). This is the single most important book in the field of embodied cognition. Here, Gibson lays out his theory of how an organism, having a particular sort of body and ways of life, perceives the opportunities and costs for acting in its environment. The book was written for an academic audience, but the writing is straightforward and assumes minimal background. If you want to know more about how to study mind from a biological perspective, then start here.

Bernd Heinrich, *Why We Run: A Natural History* (New York: Ecco, 2002). We are endurance animals and can outrun almost any other mammal over a distance of more than 20 miles in the middle of a hot sunny day. Heinrich is a highly respected biologist and award-winning nature writer, who also happens to be the 1981 national champion in the 100 k ultramarathon. If you are interested in what kind of animal we are and why we run, then Heinrich tells this story in a rich and surprising memoir.

Rachel Herz, *The Scent of Desire: Discovering Our Enigmatic Sense of Smell* (New York: William Morrow, 2007). If psychology had spent as much time and energy studying smell as it has devoted to vision, then our notions about perception and mind would be very different. You cannot experience an odor without engaging two separate processes: one identifies the smell and the other assigns it value. Roses, chocolate, and Sweetie smell wonderful. What worse insult is there then to be told "You stink"? Value is the stuff of life; and smell, more than any other sense, makes clear that things are either good or bad. Herz is one of the most eminent researchers in her field, yet she writes with an engaging ease and narrative style.

Mark Johnson, *The Meaning of the Body: Aesthetics of Human Understanding* (Chicago: University of Chicago Press, 2007). Mark Johnson was George Lakoff's philosophical collaborator for *Metaphors We Live By,* which we've referenced extensively throughout this book. *Meaning* is something of a technical, though still very accessible, modern philosopher's follow-up to *Metaphors.* Johnson beautifully argues for how the body is the route by which meaning is accessed in human life, and uses that insight to reexamine culture and the life of the mind. Of note is the final chapter, which discusses "horizontal transcendence," or what it means to have religious experience within a purely physical understanding of human experience.

John R. Krebs and Nicholas B. Davies, *An Introduction to Behavioural Ecology,* 4th ed. (Oxford: Blackwell Scientific Publications, 2012). This is the absolute best introduction to the field of behavioral ecology, which addresses questions pertaining to why animals behave the way that they do. Other than Gibson, almost no one in psychology has ever taken this approach and applied it to humans. In behavioral ecology, animals are treated as embodied, agentive, organisms that are attempting to survive, care for their offspring, and ultimately get their genes into the future. This is a textbook, but it does not read like one. The writing is lucid and the illustrations are engaging.

Bessel van der Kolk, *The Body Keeps the Score: Brain, Mind, and Body in the Healing of Trauma* (New York: Viking, 2014). The most poignant book about the body is likely Dr. van der Kolk's, where the psychiatrist first foregrounds how prevalent psychological trauma is in people's lives, and then details how it largely bypasses the language networks of the brain, necessitating nonlinguistic approaches to therapeutic practice—including the ingenious methods by which mental health professionals are leveraging embodiment to help people finally heal. *The Body Keeps the Score* is a masterwork of clinical psychology, illustrating in case study after case study the role of the body in emotional life.

NOTES

INTRODUCTION: I SING THE BODY ELECTRIC

1. C. Wedekind, T. Seebeck, F. Bettens, and A. J. Paepke. 1995. "MHC-Dependent Mate Preferences in Humans." *Proceedings of the Royal Society B: Biological Sciences* 260: 245–49. https://doi.org/10.1098/rspb.1995.0087.
2. "Butter in Japan." Butter through the Ages. http://www.webexhibits.org/butter/countries-japan.html.
3. D. J. Kelly, P. C. Quinn, A. M. Slater, K. Lee, A. Gibson, M. Smith, L. Ge, et al. 2005. "Three-Month-Olds, but Not Newborns, Prefer Own-Race Faces." *Developmental Science* 8: F31–F36. https://doi.org/10.1111/j.1467-7687.2005.0434a.x.
4. D. J. Kelly, S. Liu, L. Ge, P. C. Quinn, A. M. Slater, K. Lee, Q. Liu, et al. 2007. "Cross-Race Preferences for Same-Race Faces Extend Beyond the African Versus Caucasian Contrast in 3-Month-Old Infants." *Infancy* 11: 87–95. https://doi.org/10.1080/15250000709336871.
5. J. L. Eberhardt, P. A. Goff, V. J. Purdie, and P. G. Davies. 2004. "Seeing Black: Race, Crime, and Visual Processing." *Journal of Personality and Social Psychology* 87: 876–93. https://doi.org/10.1037/0022-3514.87.6.876.
6. M. D. Lieberman. 2017. "What Scientific Term or Concept Ought to Be More Widely Known?" *Edge*. www.edge.org/response-detail/27006.

7. J. K. Witt and D. R. Proffitt. 2005. "See the Ball, Hit the Ball: Apparent Ball Size Is Correlated with Batting Average." *Psychological Science* 16: 937–39. https://doi.org/10.1111/j.1467-9280.2005.01640.x.

8. "Mickey Mantle in His Own Words." *Aimless—with Purpose* (blog). https://hopeseguin2011.wordpress.com/2011/03/26/mickey-mantle-in-his-own-words/.

9. "George Scott Stats." 2004. Baseball Almanac. Accessed May 18, 2004. http://www.baseball-almanac.com/players/player.php?p=scottge02.

10. J. K. Witt, S. A. Linkenauger, J. Z. Bakdash, and D. R. Proffitt. 2008. "Putting to a Bigger Hole: Golf Performance Relates to Perceived Size." *Psychonomic Bulletin & Review* 15: 581–85. https://doi.org/10.3758/pbr.15.3.581.

11. J. K. Witt and T. E. Dorsch. 2009. "Kicking to Bigger Uprights: Field Goal Kicking Performance Influences Perceived Size." *Perception* 38: 1328–40. https://doi.org/10.1068/p6325.

12. Y. Lee, S. Lee, C. Carello, and M. T. Turvey. 2012. "An Archer's Perceived Form Scales the 'Hitableness' of Archery Targets." *Journal of Experimental Psychology: Human Perception and Performance* 38: 125–31. https://doi.org/10.1037/a0029036.
   R. Wesp, P. Cichello, E. B. Gracia, and K. Davis. 2004. "Observing and Engaging in Purposeful Actions with Objects Influences Estimates of Their Size." *Perception & Psychophysics* 66: 1261–67. https://doi.org/10.3758/bf03194996.

13. M. Sugovic, P. Turk, and J. K. Witt. 2016. "Perceived Distance and Obesity: It's What You Weigh, Not What You Think." *Acta Psychologica* 165: 1–8. https://doi.org/10.1016/j.actpsy.2016.01.012.

14. J. K. Witt, D. M. Schuck, and J. E. T. Taylor. 2011. "Action-Specific Effects Underwater." *Perception* 40: 530–37. https://doi.org/10.1068/p6910.

15. J. K. Witt, D. R. Proffitt, and W. Epstein. 2005. "Tool Use Affects Perceived Distance, But Only When You Intend to Use It." *Journal of Experimental Psychology: Human Perception and Performance* 31: 880–88. https://doi.org/10.1037/0096-1523.31.5.880.

16. B. Moeller, H. Zoppke, and C. Frings. 2015. "What a Car Does to Your Perception: Distance Evaluations Differ from Within and Outside of a Car." *Psychonomic Bulletin & Review* 23: 781–88. https://doi.org/10.3758/s13423-015-0954-9.

17. Andreas Kuersten. Nov. 23, 2015. "Opinion: Brain Scans in the Courtroom."

The Scientist. www.the-scientist.com/news-opinion/opinion-brain-scans-in-the-courtroom-34464.

18. Walt Whitman. "I Sing the Body Electric." Poetry Foundation. www.poetry foundation.org/poems/45472/i-sing-the-body-electric.

DOING

1. DEVELOPING

1. Karen Adolph, personal interview, June 22, 2018.

2. U. Neisser. 1981. "Obituary: James J. Gibson (1904–1979)." *American Psychologist* 36: 214–5. https://doi.org/10.1037/h0078037.

3. James J. Gibson. 1979. *The Ecological Approach to Visual Perception* (Boston: Houghton Mifflin), 127.

4. E. N. Rodkey. 2011. "The Woman Behind the Visual Cliff." *APA Monitor on Psychology* 42: 30. https://www.apa.org/monitor/2011/07-08/gibson.

5. E. J. Gibson, and R. D. Walk. 1960. "The 'Visual Cliff.'" *Scientific American* 202: 64–71. https://doi.org/10.1038/scientificamerican0460-64.

6. J. J. Campos, B. I. Bertenthal, and R. Kermoian. 1992. "Early Experience and Emotional Development: The Emergence of Wariness of Heights." *Psychological Science* 3: 61–64. https://doi.org/10.1111/j.1467-9280.1992.tb00259.x.

7. R. Held and J. Bossom. 1961. "Neonatal Deprivation and Adult Rearrangement: Complementary Techniques for Analyzing Plastic Sensory-Motor Coordinations." *Journal of Comparative and Physiological Psychology* 54: 33–37. https://doi.org/10.1037/h0046207.
   R. Held and A. Hein. 1963. "Movement-Produced Stimulation in the Development of Visually Guided Behavior." *Journal of Comparative and Physiological Psychology* 56: 872–76. https://doi.org/10.1037/h0040546.

8. Campos et al. "Early Experience and Emotional Development," 64.

9. K. E. Adolph. 2000. "Specificity of Learning: Why Infants Fall over a Veritable Cliff." *Psychological Science* 11: 290–95. https://doi.org/10.1111/1467-9280.00258.

10. Beatrix Vereijken, Karen Adolph, Mark Denny, Yaman Fadl, Simone Gill, and Ana Lucero. 1995. "Development of Infant Crawling: Balance Constraints on Interlimb Coordination," in *Studies in Perception and Action III*, ed. Benoît G. Bardy, Reinould J. Bootsma, and Yves Guiard (Mahway, NJ: Lawrence Erlbaum Associates), 255–58.

11. K. Libertus, A. S. Joh, and A. W. Needham. 2016. "Motor Training at 3 Months Affects Object Exploration 12 Months Later." *Developmental Science* 19: 1058–66. https://doi.org/10.1111/desc.12370.

    A. Needham, T. Barrett, and K. Peterman. 2002. "A Pick-me-up for Infants' Exploratory Skills: Early Simulated Experiences Reaching for Objects Using 'Sticky' Mittens Enhances Young Infants' Object Exploration Skills." *Infant Behavior and Development* 25: 279–95. https://doi.org/10.1016/S0163-6383(02)00097-8.

12. J. Faubert. 2013. "Professional Athletes Have Extraordinary Skills for Rapidly Learning Complex and Neutral Dynamic Visual Scenes." *Nature: Scientific Reports* 3: 1–3. https://doi.org/10.1038/srep01154.

13. W. Timothy Gallwey. 1974. *The Inner Game of Tennis: The Classic Guide to the Mental Side of Peak Performance* (New York: Random House; repr., Toronto: Bantam Books, 1979), 99.

14. John McEnroe with James Kaplan. 2002. *You Cannot Be Serious* (New York: G. P. Putnam's Sons; repr., London: Time Warner Paperbacks), 57.

15. J. K. Witt and M. Sugovic. 2010. "Performance and Ease Influence Perceived Speed." *Perception* 39: 1341–53. https://doi.org/10.1068/p6699.

16. R. Gray. 2013. "Being Selective at the Plate: Processing Dependence Between Perceptual Variables Relates to Hitting Goals and Performance." *Journal of Experimental Psychology: Human Perception and Performance* 39: 1124–42. https://doi.org/10.1037/a0030729.

2. WALKING

1. D. R. Proffitt, M. Bhalla, R. Gossweiler, and J. Midgett. 1995. "Perceiving Geographical Slant." *Psychonomic Bulletin & Review* 2: 409–48. https://doi.org/10.3758/BF03210980.

2. M. Bhalla and D. R. Proffitt. 1999. "Visual-Motor Recalibration in Geographical Slant Perception." *Journal of Experimental Psychology: Human Perception and Performance* 25: 1076–96. https://doi:10.1037/0096-1523.25.4.1076.

3. Ibid., 1092.

4. Kate Wong. Nov. 24, 2014. "40 Years After Lucy: The Fossil That Revolutionized the Search for Human Origins." *Observations* (blog), *Scientific American*. https://blogs.scientificamerican.com/observations/40-years-after-lucy-the-fossil-that-revolutionized-the-search-for-human-origins/.

5. Donald Johanson, personal interview, Dec. 18, 2018.

6. Ibid.

7. Charles Darwin. 1872. *The Origin of Species by Means of Natural Selection; or, The Preservation of Favored Races in the Struggle for Life and The Descent of Man and Selection in Relation to Sex,* 6th ed. (London: John Murray; repr., New York: Modern Library, 1936).

8. D. E. Lieberman. 2011. "Four Legs Good, Two Legs Fortuitous: Brains, Brawn, and the Evolution of Human Bipedalism," in *In the Light of Evolution: Essays from the Laboratory and Field,* ed. Jonathan B. Losos (Greenwood Village, CO: Roberts and Company).

9. J. A. Levine, L. M. Lanningham-Foster, S. K. McCrady, A. C. Krizan, L. R. Olson, P. H. Kane, M. D. Jensen, et al. 2005. "Interindividual Variation in Posture Allocation: Possible Role in Human Obesity." *Science* 307: 584. https://doi.org/10.1126/science.1106561.

10. Søren Kierkegaard. 2009. Letter 150, to Henriette Lund in *Kierkegaard's Writings,* vol. XXV, *Letters and Documents,* ed. and trans. Henrik Rosenmeier (Princeton, NJ: Princeton University Press), 214.

11. M. Oppezzo and D. L. Schwartz. 2014. "Give Your Ideas Some Legs: The Positive Effect of Walking on Creative Thinking." *Journal of Experimental Psychology: Learning, Memory, and Cognition* 40: 1142–52. https://doi.org/10.1037/a0036577.

12. Florence Williams. Nov. 28, 2012. "Take Two Hours of Pine Forest and Call Me in the Morning." *Outside.* Accessed Sept. 29, 2019. https://www.outsideonline.com/1870381/take-two-hours-pine-forest-and-call-me-morning.

13. Melissa Dahl. March 24, 2016. "How Running and Meditation Change the Brains of the Depressed." The Cut. https://www.thecut.com/2016/03/how-running-and-meditation-can-help-the-depressed.html.

14. Melissa Dahl. April 21, 2016. "How Neuroscientists Explain the Mind-Clearing Magic of Running." The Cut. https://www.thecut.com/2016/04/why-does-running-help-clear-your-mind.html.

15. Adie Tomer. Feb. 9, 2018. "America's Commuting Choices: 5 Major Takeaways from 2016 Census Data." *The Avenue* (blog), Brookings. www.brookings.edu/blog/the-avenue/2017/10/03/americans-commuting-choices-5-major-takeaways-from-2016-census-data/.

L. Steell, A. Garrido-Méndez, F. Petermann, X. Díaz-Martínez, M. A. Martínez,

A. M. Leiva, C. Salas-Bravo, et al. 2018. "Active Commuting Is Associated with a Lower Risk of Obesity, Diabetes, and Metabolic Syndrome in Chilean Adults." *Journal of Public Health* 40: 508–16, https://doi.org/10.1093/pubmed/fdx092.

E. Flint and S. Cummins. 2016. "Active Commuting and Obesity in Mid-Life: Cross-Sectional, Observational Evidence from UK Biobank." *The Lancet: Diabetes & Endocrinology* 4: 420–35. https://doi.org/10.1016/S2213-8587(16)00053-X.

16. Herman Pontzer. Jan. 4, 2019. "Humans Evolved to Exercise." *Scientific American*. https://www.scientificamerican.com/article/humans-evolved-to-exercise/.

17. David Attenborough. 2002. "Human Mammal, Human Hunter." *Life of Mammals*. http://www.youtube.com/watch?v=826HMLoiE_o#.

18. Christina Gough. 2018. "Number of Running Events in the U.S. from 2012 to 2016, by Distance of Race." Statista. https://www.statista.com/statistics/280485/number-of-running-events-united-states/.

19. S. Lock. 2019. "Number of Participants in Running/Jogging and Trail Running in the U.S. from 2006 to 2017." Statista. https://www.statista.com/statistics/190303/running-participants-in-the-us-since-2006/.

20. G. J. Bastien, B. Schepens, P. A. Willems, and N. C. Heglund. 2005. "Energetics of Load Carrying in Nepalese Porters." *Science* 308: 1755. https://doi.org/10.1126/science.1111513.

21. M. Sugovic, P. Turk, and J. K. Witt. 2016. "Perceived Distance and Obesity: It's What You Weigh, Not What You Think." *Acta Psychologica* 165: 1–8. https://doi.org/10.1016/j.actpsy.2016.01.012.

22. G. A. H. Taylor-Covill and F. F. Eves. 2016. "Carrying a Biological 'Backpack': Quasi-Experimental Effects of Weight Status and Body Fat Change on Perceived Steepness." *Journal of Experimental Psychology: Human Perception and Performance* 42: 331–38. http://dx.doi.org/10.1037/xhp0000137.

23. F. F. Eves. 2014. "Is There Any Proffitt in Stair Climbing? A Headcount of Studies Testing for Demographic Differences in Choice of Stairs." *Psychonomic Bulletin & Review* 21: 71–77. https://doi.org/10.3758/s13423-013-0463-7.

24. J. R. Zadra, A. L. Weltman, and D. R. Proffitt. 2016. "Walkable Distances Are Bioenergetically Scaled." *Journal of Experimental Psychology: Human Perception and Performance* 42: 39–51. https://doi.org/10.1037/xhp0000107.

25. Ibid., 49.

3. GRASPING

1. J. K. Witt and J. R. Brockmole. 2012. "Action Alters Object Identification: Wielding a Gun Increases the Bias to See Guns." *Journal of Experimental Psychology: Human Perception and Performance* 38: 1159–67. https://doi.org/10.1037/a0027881.

2. Ibid., 1166.

3. B. Kalesan, M. D. Villarreal, K. M. Keyes, and S. Galea. 2015. "Gun Ownership and Social Gun Culture." *Injury Prevention* 22: 216–20. https://doi.org/10.1136/injuryprev-2015-041586.

4. Melvyn A. Goodale and David Milner. 2004. *Sight Unseen: An Exploration of Conscious and Unconscious Vision* (Oxford: Oxford University Press).

5. Mel Goodale, personal interview, Sept. 8, 2018.

6. Ibid.

7. Ibid.

8. A. M. Haffenden, K. C. Schiff, and M. A. Goodale. 2001. "The Dissociation Between Perception and Action in the Ebbinghaus Illusion: Nonillusory Effects of Pictorial Cues on Grasp." *Current Biology* 11: 177–81. https://doi.org/10.1016/S0960-9822(01)00023-9.

9. Aristotle. 1882. "Book IV," in *On the Parts of Animals,* trans. W. Ogle. London: Kegan Paul, Trench & Co. https://archive.org/details/aristotleonparts00arisrich/page/n6.

10. Ronald Polansky. 2007. *Aristotle's De Anima: A Critical Commentary* (Cambridge: Cambridge University Press).

11. Goodale and Milner. *Sight Unseen.*

12. L. Weiskrantz, E. K. Warrington, M. D. Sanders, and J. Marshall. 1974. "Visual Capacity in the Hemianopic Field Following a Restricted Occipital Ablation." *Brain* 97: 709–28. https://doi.org/10.1093/brain/97.1.709.

13. Ibid., 726.

14. Ibid., 721.

15. B. de Gelder, M. Tamietto, G. van Boxtel, R. Goebel, A. Sahraie, J. van den Stock, B. M. C. Stienen, et al. 2008. "Intact Navigation Skills after Bilateral Loss of Striate Cortex." *Current Biology* 18: R1128–29. https://doi.org/10.1016/j.cub.2008.11.002.

16. Michael F. Land and Dan-Eric Nilsson. 2002. *Animal Eyes* (Oxford: Oxford University Press).

17. J. R. Brockmole, C. C. Davoli, R. A. Abrams, and J. K. Witt. 2013. "The World Within Reach: Effects of Hand Posture and Tool Use on Visual Cognition." *Current Directions in Psychological Science* 22: 39. https://doi.org/10.1177/0963721412465065.

18. C. L. Reed, R. Betz, J. P. Garza, and R. J. Roberts. 2010. "Grab It! Biased Attention in Functional Hand and Tool Space." *Attention, Perception, & Psychophysics* 72: 236–45. https://doi.org/10.3758/APP.72.1.236.

19. J. D. Cosman and S. P. Vecera. 2010. "Attention Affects Visual Perceptual Processing Near the Hand." *Psychological Science* 21: 1254–58. https://doi.org/10.1177/0956797610380697.
    C. C. Davoli and J. Brockmole. 2012. "The Hands Shield Attention from Visual Interference." *Attention, Perception, & Psychophysics* 74: 1386–90. https://doi.org/10.3758/s13414-012-0351-7.

20. S. A. Linkenauger, V. Ramenzoni, and D. R. Proffitt. 2010. "Illusory Shrinkage and Growth: Body-Based Rescaling Affects the Perception of Size." *Psychological Science* 21: 1318–25. https://doi.org/10.1177/0956797610380700.

21. S. A. Linkenauger, J. K. Witt, J. K. Stefanucci, J. Z. Bakdash, and D. R. Proffitt. 2009. "The Effects of Handedness and Reachability on Perceived Distance." *Journal of Experimental Psychology: Human Perception and Performance* 35: 1649–60. https://doi.org/10.1037/a0016875.

22. D. M. Abrams and M. J. Panaggio. 2012. "A Model Balancing Cooperation and Competition Can Explain Our Right-Handed World and the Dominance of Left-Handed Athletes." *Journal of the Royal Society Interface* 9: 2718–22. https://doi.org/10.1098/rsif.2012.0211.

23. M. C. Corballis. 1980. "Laterality and Myth." *American Psychologist* 35: 284–95. http://dx.doi.org/10.1037/0003-066X.35.3.284.

24. Daniel Casasanto, personal communication, March 10, 2017.

25. J. Winawer, N. Witthoft, M. C. Frank, L. Wu, A. R. Wade, and L Boroditsky. 2007. "Russian Blues Reveal the Effects of Language on Color Discriminations." *Proceedings of the National Academy of Sciences* 104: 7780–85. https://www.pnas.org/content/104/19/7780.

26. D. Casasanto and K. Dijkstra. 2010. "Motor Action and Emotional Memory." *Cognition* 115: 179–85. https://doi.org/10.1016/j.cognition.2009.11.002.

27. D. Casasanto and A. de Bruin. 2019. "Metaphors We Learn By: Directed Motor

Action Improves Word Learning." *Cognition* 182: 177–83. https://doi.org/10 .1016/j.cognition.2018.09.015.

28. Daniel Casasanto, personal communication, March 10, 2017.

29. D. Casasanto and K. Jasmin. 2010. "Good and Bad in the Hands of Politicians: Spontaneous Gestures during Positive and Negative Speech." *PLoS ONE* 14: e11805. https://doi.org/10.1371/journal.pone.0011805.

30. D. M. Oppenheimer. 2008. "The Secret Life of Fluency." *Trends in Cognitive Sciences* 12: 237–41. https://doi.org/10.1016/j.tics.2008.02.014.

31. Daniel Casasanto, personal communication, March 10, 2017.

32. D. Casasanto and E. G. Chrysikou. 2011. "When Left Is 'Right': Motor Fluency Shapes Abstract Concepts." *Psychological Science* 22: 419–22. https://doi.org /10.1177/0956797611401755.

33. Daniel Casasanto, personal communication, March 10, 2017.

KNOWING

4. THINKING

1. John Cottingham, ed. 1996. *René Descartes: Meditations on First Philosophy With Selections from the Objections and Replies,* 2nd ed. (Cambridge: Cambridge University Press), 19.

2. A. Vaccari. 2012. "Dissolving Nature: How Descartes Made us Posthuman." *Techné: Research in Philosophy and Technology* 16: 138–86. https://doi.org/10 .5840/techne201216213.

3. P. Bloom. 2007. "Religion Is Natural." *Developmental Science* 10: 147–51. https://doi.org/10.1111/j.1467-7687.2007.00577.x.

4. P. Bloom. 2006. "My Brain Made Me Do It." *Journal of Cognition and Culture* 6: 209–14. https://doi.org/10.1163/156853706776931303.

5. N. Kandasamy, S. N. Garfinkel, L. Page, B. Hardy, H. D. Critchley, M. Gurnell, and J. M. Coates. 2016. "Interoceptive Ability Predicts Survival on a London Trading Floor." *Scientific Reports* 6: 1–7. https://doi.org/10.1038 /srep32986.

6. Ibid., 5.

7. Sarah Garfinkel, personal communication, Jan. 19, 2017.

8. D. Fischer, M. Messner, and O. Pollatos. 2017. "Improvement of Interoceptive

Processes after an 8-Week Body Scan Intervention." *Frontiers in Human Neuroscience* 11: 452. https://doi.org/10.3389/fnhum.2017.00452.

9. L. G. Kiken, N. J. Shook, J. L. Robins, and J. N. Clore. 2018. "Association between Mindfulness and Interoceptive Accuracy in Patients with Diabetes: Preliminary Evidence from Blood Glucose Estimates." *Complementary Therapies in Medicine* 36: 90–92. https://doi.org/10.1016/j.ctim.2017.12.003.

10. X. T. Wang and R. D. Dvorak. 2010. "Sweet Future: Fluctuating Blood Glucose Levels Affect Future Discounting." *Psychological Science* 21: 183–88. https://doi.org/10.1177/0956797609358096.

11. C. N. DeWall, R. F. Baumeister, M. T. Gailliot, and J. K. Maner. 2008. "Depletion Makes the Heart Grow Less Helpful: Helping as a Function of Self-Regulatory Energy and Genetic Relatedness." *Personality and Social Psychology Bulletin* 34: 1653–62. https://doi.org/10.1177/0146167208323981.

12. S. Danziger, J. Levav, and L. Avnaim-Pesso. 2011. "Extraneous Factors in Judicial Decisions." *Proceedings of the National Academy of Sciences* 108: 6889–92. https://doi.org/10.1073/pnas.1018033108.

13. M. L. Anderson, J. Gallagher, and E. Ramirez Ritchie. May 3, 2017. "How the Quality of School Lunch Affects Students' Academic Performance." *Brown Center Chalkboard* (blog). Brookings. Accessed June 30, 2019. https://www.brookings.edu/blog/brown-center-chalkboard/2017/05/03/how-the-quality-of-school-lunch-affects-students-academic-performance/.

14. Jane E. Brody. June 5, 2017. "Feeding Young Minds: The Importance of School Lunches." *New York Times*. https://www.nytimes.com/2017/06/05/well/feeding-young-minds-the-importance-of-school-lunches.html.

15. "Facts About Child Nutrition." n.d. National Education Association. http://www.nea.org/home/39282.htm.

16. N. Schwarz, H. Bless, F. Strack, G. Klumpp, H. Rittenauer-Schatka, and A. Simons. 1991. "Ease of Retrieval as Information: Another Look at the Availability Heuristic." *Journal of Personality and Social Psychology* 61: 195–202. https://doi.org/10.1037/0022-3514.61.2.195.

17. M. S. McGlone and J. Tofighbakhsh. 1999. "The Keats Heuristic: Rhyme as Reason in Aphorism Interpretation." *Poetics* 26: 235–44. https://doi.org/10.1016/S0304-422X(99)00003-0.

18. Friedrich Nietzsche. 2001. *The Gay Science: With a Prelude in German Rhymes and an Appendix of Songs,* ed. Bernard Williams, trans. Josefine Nauckhoff, poems trans. Adrian Del Caro (Cambridge: Cambridge University Press).

19. Daniel Kahneman, Paul Slovic, and Amos Tversky, eds. *Judgement under Uncertainty: Heuristics and Biases* (Cambridge: Cambridge University Press).

20. A. Nowrasteh. 2016. "Terrorism and Immigration: A Risk Analysis." *Cato Institute Policy Analysis No. 798,* 1–26. https://papers.ssrn.com/sol3/papers.cfm?abstract_id=2842277.

21. Dave Mosher and Skye Gould. Jan. 31, 2017. "How Likely Are Foreign Terrorists to Kill Americans? The Odds May Surprise You." Business Insider. Accessed Nov. 11, 2019. https://www.businessinsider.com/death-risk-statistics-terrorism-disease-accidents-2017-1.

22. Drake Baer. Jan. 12, 2017. "The Shows You Watch Build Your Perception of the World." The Cut. https://www.thecut.com/2017/01/the-shows-you-watch-build-your-perception-of-the-world.html.

23. S. Murrar and M. Brauer. 2017. "Entertainment-Education Effectively Reduces Prejudice." *Group Processes & Intergroup Relations* 21: 1053–77. https://doi.org/10.1177/1368430216682350.

24. Ryan Boyd, personal communication, June 9, 2017.

25. S. R. Sommers. 2006. "On Racial Diversity and Group Decision Making: Identifying Multiple Effects of Racial Composition on Jury Deliberations." *Journal of Personality and Social Psychology* 90: 597–612. https://doi.org/10.1037/0022-3514.90.4.597.

26. Courtroom Television Network. Jan. 1995. "*Georgia v. Redding*: A Rapist on Trial: DNA Takes the Stand."

27. Sommers. "On Racial Diversity and Group Decision Making," 607.

28. Ibid., 608.

29. Harry G. Frankfurt. 2005. *On Bullshit* (Princeton, NJ: Princeton University Press).

30. A. Perrin. Oct. 2015. "Social Media Usage: 2005–2015." Pew Research Center Internet & Technology. https://www.pewinternet.org/2015/10/08/social-networking-usage-2005-2015/.

31. Jane Meyer. Sept. 24, 2018. "How Russia Helped Swing the Election for Trump." *New Yorker.* Accessed June 30, 2019. https://www.newyorker.com/magazine/2018/10/01/how-russia-helped-to-swing-the-election-for-trump.

32. J. V. Petrocelli. 2018. "Antecedents of Bullshitting." *Journal of Experimental Social Psychology* 76: 249–58. https://doi.org/10.1016/j.jesp.2018.03.004.

33. F. T. Bacon. 1979. "Credibility of Repeated Statements: Memory for Trivia." *Journal of Experimental Psychology: Human Learning and Memory* 5: 241–52. https://doi.org/10.1037/0278-7393.5.3.241.

34. L. K. Fazio, N. M. Brashier, B. K. Payne, and E. J. Marsh. 2015. "Knowledge Does Not Protect Against Illusory Truth." *Journal of Experimental Psychology: General* 144: 993–1002. https://doi.org/10.1037/xge0000098.

5. FEELING

1. Y. Inbar, D. A. Pizarro, and P. Bloom. 2009. "Conservatives Are More Easily Disgusted Than Liberals." *Cognition and Emotion* 23: 725. https://doi.org/10.1080/02699930802110007.

2. Ibid.

3. J. T. Crawford, Y. Inbar, and V. Maloney. 2014. "Disgust Sensitivity Selectively Predicts Attitudes Toward Groups That Threaten (or Uphold) Traditional Sexual Morality." *Personality and Individual Differences* 70: 218–23. https://doi.org/10.1016/j.paid.2014.07.001.

4. Y. Inbar, D. A. Pizarro, and P. Bloom. 2012. "Disgusting Smells Cause Decreased Liking of Gay Men." *Emotion* 12: 23–27. https://doi.org/10.1037/a0023984.

5. W-Y. Ahn, K. T. Kishida, X. Gu, T. Lohrenz, A. Harvey, J. R. Alford, K. B. Smith, et al. 2014. "Nonpolitical Images Evoke Neural Predictors of Political Ideology." *Current Biology* 24: 2693–99. https://doi.org/10.1016/j.cub.2014.09.050.

6. David Hume. 1739–1740. Book 3, part 1, Sect. II: "Moral Distinctions Derived from a Moral Sense," in *A Treatise of Human Nature*. https://davidhume.org/texts/t/3/1/2.

7. Gerald L. Clore. 2018. "The Impact of Affect Depends on its Object," in *The Nature of Emotion: Fundamental Questions*, ed. Andrew S. Fox, Regina C. Lapate, Alexander J. Shackman, and Richard J. Davidson, 2nd ed. (Oxford: Oxford University Press), 188–89.

8. Merriam-Webster. "imperative." https://www.merriam-webster.com/dictionary/imperative.

9. Colin Klein. 2015. *What the Body Commands: The Imperative Theory of Pain* (Cambridge, MA: MIT Press), 4.

10. F. B. Axelrod and G. Gold-von Simson. 2007. "Hereditary Sensory and Autonomic Neuropathies: Types II, III, and IV." *Orphanet Journal of Rare Diseases* 2: 1–12. https://doi.org/10.1186/1750-1172-2-39.

11. P. Rainville. 2002. "Brain Mechanisms of Pain Affect and Pain Modulation." *Current Opinion in Neurobiology* 12: 195–204. https://doi.org/10.1016/S0959 -4388(02)00313-6.

12. G. MacDonald and M. R. Leary. 2005. "Why Does Social Exclusion Hurt? The Relationship Between Social and Physical Pain." *Psychological Bulletin* 131: 202–23. https://doi.org/10.1037/0033-2909.131.2.202.

13. J. Panksepp, B. Herman, R. Conner, P. Bishop, and J. P. Scott. 1978. "The Biology of Social Attachments: Opiates Alleviate Separation Distress." *Biological Psychiatry* 13: 607–18.

14. N. I. Eisenberger, M. D. Lieberman, and K. D. Williams. 2003. "Does Rejection Hurt? An fMRI Study of Social Exclusion." *Science* 302: 290–92. https:// doi.org/10.1126/science.1089134.

15. J. B. Silk, S. C. Alberts, and J. Altmann. 2003. "Social Bonds of Female Baboons Enhance Infant Survival." *Science* 302: 1231–34. https://doi.org/10.1126 /science.1088580.

16. N. Schwarz and G. L. Clore. 1983. "Mood, Misattribution, and Judgments of Well-Being: Informative and Directive Functions of Affective States." *Journal of Personality and Social Psychology* 45: 519–23. https://psycnet.apa.org/doi/10 .1037/0022-3514.45.3.513.

17. Ibid., 519.

18. World Health Organization. 2017. *Depression and Other Common Mental Disorders: Global Health Estimates.* Geneva: World Health Organization.

19. G. L. Clore and J. Palmer. 2009. "Affective Guidance of Intelligent Agents: How Emotion Controls Cognition." *Cognitive Systems Research* 10: 21–30. https:// doi.org/10.1016/j.cogsys.2008.03.002.

20. E. R. Watkins. 2008. "Constructive and Unconstructive Repetitive Thought." *Psychological Bulletin* 134: 163–206. https://doi.org/10.1037/0033-2909.134.2 .163.

21. S. Nolen-Hoeksema. 2000. "The Role of Rumination in Depressive Disorders

and Mixed Anxiety/Depressive Symptoms." *Journal of Abnormal Psychology* 109: 504–11. https://doi.org/10.1037/0021-843X.109.3.504.

J. Spasojević and L. B. Alloy. 2001. "Rumination as a Common Mechanism Relating Depressive Risk Factors to Depression." *Emotion* 1: 25–37. https://doi.org/10.1037//1528-3542.1.1.25.

22. E. Watkins, N. J. Moberly, and M. L. Moulds. 2008. "Processing Mode Causally Influences Emotional Reactivity: Distinct Effects of Abstract Versus Concrete Construal on Emotional Response." *Emotion* 8: 364–78. https://doi.org/10.1037/1528-3542.8.3.364.

23. Drake Baer. Oct. 16, 2017. "How Psychologists Change the Mental Habits That Drive Depression." Thrive Global. https://thriveglobal.com/stories/retraining-mental-habits-of-depression/.

24. S. Schnall, J. Haidt, G. L. Clore, and A. H. Jordan. 2008. "Disgust as Embodied Moral Judgment." *Personality and Social Psychology Bulletin* 34: 1096–1109. https://doi.org/10.1177/0146167208317771.

25. Laura D'Olimpio. June 2, 2016. "The Trolley Dilemma: Would You Kill One Person to Save Five?" *The Conversation*. http://theconversation.com/the-trolley-dilemma-would-you-kill-one-person-to-save-five-57111.

26. J. K. Stefanucci and J. Storbeck. 2009. "Don't Look Down: Emotional Arousal Elevates Height Perception." *Journal of Experimental Psychology: General* 138: 131–45. https://doi.org/10.1037/a0014797.

27. Ibid.

28. E. A. Phelps, S. Ling, and M. Carrasco. 2006. "Emotion Facilitates Perception and Potentiates the Perceptual Benefits of Attention." *Psychological Science* 17: 292–99. https://doi.org/10.1111/j.1467-9280.2006.01701.x.

29. J. K. Stefanucci, D. R. Proffitt, G. L. Clore, and N. Parekh. 2008. "Skating down a Steeper Slope: Fear Influences the Perception of Geographical Slant." *Perception* 37: 321–23. https://doi.org/10.1068/p5796.

30. S. Rachman and M. Cuk. 1992. "Fearful Distortions." *Behaviour Research and Therapy* 30: 583. https://doi.org/10.1016/0005-7967(92)90003-Y.

31. C. R. Riener, J. K. Stefanucci, D. R. Proffitt, and G. Clore. 2011. "An Effect of Mood on the Perception of Geographical Slant." *Cognition & Emotion* 25: 174–82. https://doi.org/10.1080/02699931003738026.

32. Clore and Palmer. "Affective Guidance of Intelligent Agents."

33. T. Leibovich, N. Cohen, and A. Henik. 2016. "Itsy bitsy spider?: Valence and

Self-Relevance Predict Size Estimation." *Biological Psychology* 121: 138–45. https://doi.org/10.1016/j.biopsycho.2016.01.009.

M. W. Vasey, M. R. Vilensky, J. H. Heath, C. N. Harbaugh, A. G. Buffington, and R. H. Fazio. 2012. "It was as big as my head, I swear!: Biased Spider Size Estimation in Spider Phobia." *Journal of Anxiety Disorders* 26: 20–24. https://doi.org/10.1016/j.janxdis.2011.08.009.

34. J. K. Witt and M. Sugovic. 2013. "Spiders Appear to Move Faster Than Non-Threatening Objects Regardless of One's Ability to Block Them." *Acta Psychologica* 143: 284–91. https://doi.org/10.1016/j.actpsy.2013.04.011.

35. S. Cole, E. Balcetis, and D. Dunning. 2012. "Affective Signals of Threat Increase Perceived Proximity." *Psychological Science* 24: 34–40. https://doi.org/10.1177/0956797612446953.

36. Drake Baer. Sept. 17, 2004. "This Stanford Psychologist Won a MacArthur Genius Grant for Showing How Unconsciously Racist Everybody Is." Business Insider. https://www.businessinsider.com/stanford-psychologist-macarthur-genius-on-racism-2014-9?international=true&r=US&IR=T.

37. MacArthur Foundation. Sept. 16, 2014. "Social Psychologist Jennifer L. Eberhardt, 2014 MacArthur Fellow." Accessed Oct. 25, 2019. https://www.youtube.com/watch?v=lsV8kiDtN78.

38. J. L. Eberhardt, P. G. Davies, V. J. Purdie-Vaughns, and S. L. Johnson. 2006. "Looking Deathworthy: Perceived Stereotypicality of Black Defendants Predicts Capital-Sentencing Outcomes." *Psychological Science* 17: 383–86. https://doi.org/10.1111/j.1467-9280.2006.01716.x.

39. J. Correll, B. Park, C. M. Judd, and B. Wittenbrink. 2007. "The Influence of Stereotypes on Decisions to Shoot." *European Journal of Social Psychology* 37: 1102–17. https://doi.org/10.1002/ejsp.450.

6. SPEAKING

1. A. Kendon. 2017. "Reflections on the 'Gesture-First' Hypothesis of Language Origins." *Psychonomic Bulletin & Review* 24: 163–70. https://doi.org/10.3758/s13423-016-1117-3.

2. G. Króliczak, B. J. Piper, and S. H. Frey. 2011. "Atypical Lateralization of Language Predicts Cerebral Asymmetries in Parietal Gesture Representations." *Neuropsychologia* 49: 1698–1702. https://doi.org/10.1016/j.neuropsychologia.2011.02.044.

3. L. Vainio, M. Schulman, K. Tiippana, and M. Vainio. 2013. "Effect of Syllable Articulation on Precision and Power Grip Performance." *PLoS ONE* 8: e53061. https://doi.org/10.1371/journal.pone.0053061.

4. John J. Ohala. 1994. "The Frequency Code Underlies the Sound-Symbolic Use of Voice Pitch," in *Sound Symbolism,* ed. Leanne Hinton, Johanna Nichols, and John J. Ohala (Cambridge: Cambridge University Press), 325–47.

5. D. S. Schmidtke, M. Conrad, and A. M. Jacobs. Feb. 12, 2014. "Phonological Iconicity." *Frontiers in Psychology* 5. https://doi.org/10.3389/fpsyg.2014 .00080.

6. Anassa Rhenisch. Feb. 18, 2015. "Phonesthemes." Accessed Nov. 3, 2019. https://anassarhenisch.wordpress.com/2015/02/18/phonesthemes/.

7. J. M. Iverson and S. Goldin-Meadow. 1998. "Why People Gesture When They Speak." *Nature* 396: 228. https://doi.org/10.1038/24300.

8. Ibid.

9. Ibid.

10. Drake Baer. July 5, 2016. "Talking with Your Hands Makes You Learn Things Faster." The Cut. https://www.thecut.com/2016/07/talking-with-your-hands -makes-you-learn-things-faster.html.

11. R. M. Ping, S. Goldin-Meadow, and S. L. Beilock. 2014. "Understanding Gesture: Is the Listener's Motor System Involved?" *Journal of Experimental Psychology: General* 143: 196. https://doi.org/10.1037/a0032246.

12. Susan Goldin-Meadow, personal communication, June 29, 2016.

13. David McNeill. 1996. *Hand and Mind: What Gestures Reveal about Thought,* new ed. (Chicago: University of Chicago Press).

14. S. Goldin-Meadow. 1997. "When Gestures and Words Speak Differently." *Current Directions in Psychological Science* 6: 138–43. https://doi.org/10.1111/1467 -8721.ep10772905.

15. E. S. LeBarton, S. Goldin-Meadow, and S. Raudenbush. 2015. "Experimentally Induced Increases in Early Gesture Lead to Increases in Spoken Vocabulary." *Journal of Cognition and Development* 16: 199–220. https://doi.org/10 .1080/15248372.2013.858041.

16. M. A. Novack, E. L. Congdon, N. Hemani-Lopez, and S. Goldin-Meadow. 2014. "From Action to Abstraction: Using the Hands to Learn Math." *Psychological Science* 25: 903–10. https://doi.org/10.1177/0956797613518351.

17. Nicholas Day. March 26, 2013. "How Pointing Makes Babies Human," *How*

Babies Work (blog). *Slate.* https://slate.com/human-interest/2013/03/research
-on-babies-and-pointing-reveals-the-actions-importance.html?via=recirc
_recent.

18. J. M. Iverson and S. Goldin-Meadow. 2005. "Gesture Paves the Way for Language Development." *Psychological Science* 16: 367–71. https://doi.org/10.1111
/j.0956-7976.2005.01542.x.

19. Patricia Marks Greenfield and Joshua H. Smith. 1977. *The Structure of Communication in Early Language Development* (New York: Academic Press), xi, 238.

20. M. Tomasello, B. Hare, H. Lehmann, and J. Call. 2007. "Reliance on Head versus Eyes in the Gaze Following of Great Apes and Human Infants: The Cooperative Eye Hypothesis." *Journal of Human Evolution* 52: 314–20. https://doi
.org/10.1016/j.jhevol.2006.10.001.

21. A. L. Ferry, S. J. Hespos, and S. R. Waxman. 2010. "Categorization in 3- and 4-Month-Old Infants: An Advantage of Words Over Tones." *Child Development* 81: 472–79. https://doi.org/10.1111/j.1467-8624.2009.01408.x.

22. Merriam-Webster. "taxicab." https://www.merriam-webster.com/dictionary
/taxicab.

23. S. Harnad. 1990. "The Symbol Grounding Problem." *Physica D: Nonlinear Phenomena* 42: 335–46. https://doi.org/10.1016/0167-2789(90)90087-6.

24. A. M. Glenberg, M. Sato, L. Cattaneo, L. Riggio, D. Palumbo, and G. Buccino. 2008. "Processing Abstract Language Modulates Motor System Activity." *Quarterly Journal of Experimental Psychology* 61: 905–19. https://doi.org/10
.1080/17470210701625550.

25. O. Hauk, I. Johnsrude, and F. Pulvermüller. 2004. "Somatotopic Representation of Action Words in Human Motor and Premotor Cortex." *Neuron* 41: 301–7. https://doi.org/10.1016/S0896-6273(03)00838-9.

26. F. Carota, R. Moseley, and F. Pulvermüller. 2012. "Body-Part-Specific Representations of Semantic Noun Categories." *Journal of Cognitive Neuroscience* 24: 1492–1509. https://doi.org/10.1162/jocn_a_00219.

27. R. M. Willems, I. Toni, P. Hagoort, and D. Casasanto. 2009. "Body-Specific Motor Imagery of Hand Actions: Neural Evidence from Right-and Left-Handers." *Frontiers in Human Neuroscience* 3. https://doi.org/10.3389/neuro.09.039
.2009.

28. F. R. Dreyer and F. Pulvermüller. 2018. "Abstract Semantics in the Motor

System?—An Event-Related fMRI Study on Passive Reading of Semantic Word Categories Carrying Abstract Emotional and Mental Meaning." *Cortex* 100: 52–70. https://doi.org/10.1016/j.cortex.2017.10.021.

29. D. A. Havas, A. M. Glenberg, K. A. Gutowski, M. J. Lucarelli, and R. J. David-son. 2010. "Cosmetic Use of Botulinum Toxin-A Affects Processing of Emo-tional Language." *Psychological Science* 21: 895–900. https://doi.org/10.1177 /0956797610374742.

30. "Challenges to Botox Threaten a Market Makeover." March 8, 2018. *Financial Times*. Accessed June 30, 2019. https://www.ft.com/content/49570b38-221f -11e8-9a70-08f715791301.

31. Centers for Disease Control and Prevention. "About Botulism." Botulism. Ac-cessed June 30, 2019. https://www.cdc.gov/botulism/general.html.

32. G. Defazio, G. Abbruzzese, P. Girlanda, L. Vacca, A. Currà, R. De Salvia et al. 2002. "Botulinum Toxin A Treatment for Primary Hemifacial Spasm." *Ar-chives of Neurology* 59: 418–20. https://doi.org/10.1001/archneur.59.3.418.

M. Khalil, H. W. Zafar, V. Quarshie, and F. Ahmed. 2014. "Prospective Analy-sis of the Use of OnabotulinumtoxinA (BOTOX) in the Treatment of Chronic Migraine; Real-life Data in 254 Patients from Hull, UK." *Journal of Headache and Pain* 15: 1–9. https://doi.org/10.1186/1129-2377-15-54.

R. Gooriah and F. Ahmed. 2015. "OnabotulinumtoxinA for Chronic Migraine: A Critical Appraisal." *Therapeutics and Clinical Risk Management* 11: 1003–13. https://doi.org/10.2147/TCRM.S76964.

33. J.-C. Baumeister, G. Papa, and F. Foroni. 2016. "Deeper than Skin Deep—The Effect of Botulinum Toxin-A on Emotion Processing." *Toxicon* 118: 86–90. https://doi.org/10.1016/j.toxicon.2016.04.044.

34. E. Santana and M. de Vega. 2011. "Metaphors Are Embodied, and So Are Their Literal Counterparts." *Frontiers in Psychology* 2. https://doi.org/10.3389 /fpsyg.2011.00090.

35. A. M. Glenberg. 2011. "How Reading Comprehension Is Embodied and Why That Matters." *International Electronic Journal of Elementary Education* 4: 5–18. https://iejee.com/index.php/IEJEE/article/view/210.

36. Ibid., 15.

37. I. Walker and C. Hulme. 1999. "Concrete Words Are Easier to Recall Than Abstract Words: Evidence for a Semantic Contribution to Short-Term Serial

Recall." *Journal of Experimental Psychology: Learning, Memory, and Cognition* 25: 1256–71. https://doi.org/10.1037/0278-7393.25.5.1256.

38. E. Jefferies, K. Patterson, R. W. Jones, and M. A. Lambon Ralph. 2009. "Comprehension of Concrete and Abstract Words in Semantic Dementia." *Neuropsychology* 23: 492–99. https://doi.org/10.1037/a0015452.

39. Steven Pinker. 2014. *The Sense of Style: The Thinking Person's Guide to Writing in the 21st Century* (London: Penguin Books).

40. Drake Baer. Feb. 16, 2016. "I've Been Listening to Almost Nothing but Taylor Swift for 3 Weeks and I'm Convinced It's Made Me a Better Writer." Business Insider. https://www.businessinsider.com/how-taylor-swift-made-me-a-better-writer-2016-2.

41. Ludwig Wittgenstein. 1953. *Philosophical Investigations,* trans. G. E. M. Anscombe (Oxford: Blackwell), 224.

BELONGING

7. CONNECTING

1. R. A. Spitz. 1945. "Hospitalism: An Inquiry into the Genesis of Psychiatric Conditions in Early Childhood." *The Psychoanalytic Study of the Child* 1: 53–74. https://doi.org/10.1080/00797308.1945.11823126.

2. Floyd M. Crandall. 1897. "Hospitalism." *Archives of Pediatrics* 14: 448–54. Neonatology on the Web. http://www.neonatology.org/classics/crandall.html.

3. Thomas E. Cone, Jr. 1980. "Perspectives in Neonatology," in *Historical Review and Recent Advances in Neonatal and Perinatal Medicine,* ed. George F. Smith and Dharmapuri Vidyasagar (Mead Johnson Nutritional Division), 1. Neonatology on the Web. http://www.neonatology.org/classics/mj1980/ch02.html.

4. Robert Karen. 1994. *Becoming Attached: First Relationships and How They Shape Our Capacity to Love* (New York: Warner Books; repr., Oxford: Oxford University Press, 1998).

5. Ibid., 25.

6. L. F. Berkman and S. L. Syme. 1979. "Social Networks, Host Resistance, and Mortality: A Nine-Year Follow-up Study of Alameda County Residents." *American Journal of Epidemiology* 109: 186–204. https://doi.org/10.1093/oxfordjournals.aje.a112674.

7. K. A. Ertel, M. M. Glymour, and L. F. Berkman. 2009. "Social Networks and Health: A Life Course Perspective Integrating Observational and Experimental Evidence." *Journal of Social and Personal Relationships* 26: 73–92. https://doi .org/10.1177/0265407509105523.

S. O. Roper and J. B. Yorgason. 2009. "Older Adults with Diabetes and Osteoarthritis and Their Spouses: Effects of Activity Limitations, Marital Happiness, and Social Contacts on Partners' Daily Mood." *Family Relations* 58: 460–74. https://doi.org/10.1111/j.1741-3729.2009.00566.x.

J. B. Yorgason, S. O. Roper, J. G. Sandberg, and C. A. Berg. 2012. "Stress Spillover of Health Symptoms from Healthy Spouses to Patient Spouses in Older Married Couples Managing Both Diabetes and Osteoarthritis." *Families, Systems, & Health* 30: 330–43. https://doi.org/10.1037/a0030670.

B. N. Uchino. 2006. "Social Support and Health: A Review of Physiological Processes Potentially Underlying Links to Disease Outcomes." *Journal of Behavioral Medicine* 29: 377–87. https://doi.org/10.1007/s10865-006-9056-5.

Lily Dayton. Sept. 13. 2013. "Social Support Network May Add to Longevity." *Los Angeles Times.* https://www.latimes.com/health/la-xpm-2010-sep-13-la-he -friends-health-20100913-story.html.

8. J. Holt-Lunstad, T. B. Smith, and J. B. Layton. 2010. "Social Relationships and Mortality Risk: A Meta-analytic Review." *PLoS Medicine:* e1000316. https://doi .org/10.1371/journal.pmed.1000316.

9. Selby Frame. Oct. 18, 2017. "Julianne Holt-Lunstad Probes Loneliness, Social Connections." American Psychological Association. Accessed June 30, 2019. https://www.apa.org/members/content/holt-lunstad-loneliness-social -connections.

10. David Abulafia. Jan. 24, 2014. "'Inventing the Individual,' by Larry Siedentop." *Financial Times.* Accessed June 30, 2019. https://www.ft.com/content /26722be8-81f1-11e3-87d5-00144feab7de.

11. "individualism." n.d. Online Etymology Dictionary. https://www.etymonline .com/word/individualism. Accessed Feb. 9, 2020.

12. H. F. Harlow. 1958. "The Nature of Love." *American Psychologist* 13: 673–85.

13. Ibid., 677.

14. I. Morrison, L. S. Löken, and H. Olausson. 2010. "The Skin as a Social Organ." *Experimental Brain Research* 204: 305–14. https://doi.org/10.1007 /s00221-009-2007-y.

15. J. Lehmann, A. H. Korstjens, and R. I. M. Dunbar. 2007. "Group Size, Grooming and Social Cohesion in Primates." *Animal Behavior* 74: 1617–29. https://doi.org/10.1016/j.anbehav.2006.10.025.

16. H. Olausson, Y. Lamarre, H. Backlund, C. Morin, B. G. Wallin, G. Starck, S. Ekholm, et al. 2002. "Unmyelinated Tactile Afferents Signal Touch and Project to Insular Cortex." *Nature Neuroscience* 5: 900–4. https://doi.org/10.1038/nn896.

17. Jim Coan, personal interview, July 18, 2017.

18. J. A. Coan, L. Beckes, M. Z. Gonzalez, E. L. Maresh, C. L. Brown, and K. Hasselmo. 2017. "Relationship Status and Perceived Support in the Social Regulation of Neural Responses to Threat." *Social Cognitive and Affective Neuroscience* 12: 1574–83. https://doi.org/10.1093/scan/nsx091.

19. J. A. Coan and D. A. Sbarra. 2015. "*Social Baseline Theory*: The Social Regulation of Risk and Effort." *Current Opinion in Psychology* 1: 87–91. https://doi.org/10.1016/j.copsyc.2014.12.021.

20. Coan et al. "Relationship Status and Perceived Support in the Social Regulation of Neural Responses to Threat." p. 1580

21. A. Doerrfeld, N. Sebanz, and M. Shiffrar. 2012. "Expecting to Lift a Box Together Makes the Load Look Lighter." *Psychological Research* 76: 467–75. https://doi.org/10.1007/s00426-011-0398-4.

22. S. Schnall, K. D. Harber, J. K. Stefanucci, and D. R. Proffitt. 2008. "Social Support and the Perception of Geographical Slant." *Journal of Experimental Social Psychology* 44: 1246–55. https://doi.org/10.1016/j.jesp.2008.04.011.

23. Joel Goldberg. July 30, 2016. "It Takes a Village to Determine the Origins of an African Proverb." *Goats and Soda: Stories of Life in a Changing World* (blog). NPR. https://www.npr.org/sections/goatsandsoda/2016/07/30/487925796/it-takes-a-village-to-determine-the-origins-of-an-african-proverb.

24. Sarah Blaffer Hrdy. 2009. *Mothers and Others: The Evolutionary Origins of Mutual Understanding* (Cambridge, MA: Harvard University Press).

25. Sarah Blaffer Hrdy. 2005. "Comes the Child before Man: How Cooperative Breeding and Prolonged Postweaning Dependence Shaped Human Potential," in *Hunter-Gatherer Childhoods: Evolutionary, Developmental, and Cultural Perspectives,* ed. Barry S. Hewlett and Michael E. Lamb (New York: Routledge), 65–91.

26. Graham Townsley. Oct. 26, 2009. "Challenging a Paradigm." *Evolution* (blog). NOVA. https://www.pbs.org/wgbh/nova/article/evolution-motherhood/.

27. A. C. Kruger and M. Konner. 2010. "Who Responds to Crying?: Maternal Care and Allocare among the !Kung." *Human Nature* 21: 309–29. https://doi.org/10.1007/s12110-010-9095-z.

28. Sarah B. Hrdy. 2016. "Development Plus Social Selection in the Emergence of 'Emotionally Modern' Humans," in *Childhood: Origins, Evolution, and Implications,* ed. Courtney L. Meehan and Alyssa A. Crittenden (Albuquerque: University of New Mexico Press), 12.

29. M. L. Kringelbach, A. Lehtonen, S. Squire, A. G. Harvey, M. G. Craske, I. E. Holliday, A. L. Green, et al. 2008. "A Specific and Rapid Neural Signature for Parental Instinct." *PloS ONE* 3: e1664. https://doi.org/10.1371/journal.pone.0001664.

30. M. L. Glocker, D. D. Langleben, K. Ruparel, J. W. Loughead, R. C. Gur, and N. Sachser. 2009. "Baby Schema in Infant Faces Induces Cuteness Perception and Motivation for Caretaking in Adults." *Ethology* 115: 257–63. https://doi.org/10.1111/j.1439-0310.2008.01603.x.

31. Drake Baer. June 23, 2016. "France Has More Babies Than Everybody in Europe Because of Day Care and Prussia." The Cut. https://www.thecut.com/2016/06/france-has-more-babies-than-everybody-in-europe.html.
Ian Centrone. July 3, 2019. "Japan Has Too Many Abandoned Schools—So They're Turning Them into Community Centers and Aquariums." *Travel + Leisure*. https://www.travelandleisure.com/culture-design/abandoned-schools-in-japan-transformed-into-cultural-centers.

32. Gretchen Livingston. Aug. 8, 2019. "Hispanic Women No Longer Account for the Majority of Immigrant Births in the U.S." Fact Tank: News in the Numbers. Pew Research Center. https://www.pewresearch.org/fact-tank/2019/08/08/hispanic-women-no-longer-account-for-the-majority-of-immigrant-births-in-the-u-s/.

33. Karen. *Becoming Attached*, 89.

34. K. Lorenz. 1935. *"Der Kumpan in der Umwelt des Vogels. Der Artgenosse als auslösendes Moment sozialer Verhaltensweisen." Journal für Ornithologie* 83: 137–215, 289–413.

35. Karen. *Becoming Attached*, 90.

36. Drake Baer. Nov. 17, 2016. "This Revolutionary Parenting Insight Will

Help Your Love Life." The Cut. https://www.thecut.com/2016/11/why-people
-project-their-parents-on-their-partners.html.

37. Ibid.

38. T. Ein-Dor and O. Tal. 2012. "Scared Saviors: Evidence that People High in Attachment Anxiety Are More Effective in Alerting Others to Threat." *European
Journal of Social Psychology* 42: 667–71. https://doi.org/10.1002/ejsp.1895.

39. Drake Baer. Oct. 20, 2017. "This is How You Raise Successful Teens." Thrive
Global. https://thriveglobal.com/stories/this-is-how-you-raise-successful-teens/.

40. D. A. Bennett, J. A. Schneider, Y. Tang, S. E. Arnold, and R. S. Wilson. 2006.
"The Effect of Social Networks on the Relation between Alzheimer's Disease
Pathology and Level of Cognitive Function in Old People: A Longitudinal Cohort Study." *The Lancet Neurology* 5: 406–12. https://doi.org/10.1016/S1474
-4422(06)70417-3.

8. IDENTIFYING

1. David Brooks. July 15, 2016. "We Take Care of Our Own." *New York Times*.
Accessed Dec. 4, 2019. https://www.nytimes.com/2016/07/15/opinion/we-take
-care-of-our-own.html.

2. H. Rusch. 2014. "The Evolutionary Interplay of Intergroup Conflict and Altruism in Humans: A Review of Parochial Altruism Theory and Prospects
for Its Extension." *Proceedings of the Royal Society B: Biological Sciences* 281:
20141539. https://doi.org/10.1098/rspb.2014.1539.

3. Brian Handwerk. Jan. 20, 2016. "An Ancient, Brutal Massacre May Be
the Earliest Evidence of War." *Smithsonian*. https://www.smithsonianmag
.com/science-nature/ancient-brutal-massacre-may-be-earliest-evidence-war
-180957884/.

4. Milton Leitenberg. 2006. "Deaths in Wars and Conflicts in the 20th Century."
3rd ed. Cornell University Peace Studies Program. Occasional paper 29. https://
www.clingendael.org/sites/default/files/pdfs/20060800_cdsp_occ_leitenberg
.pdf.

5. V. Martínez-Tur, V. Peñarroja, M. A. Serrano, V. Hidalgo, C. Moliner, A. Sal-
vador, A Alacreu-Crespo, et al. 2014. "Intergroup Conflict and Rational Decision Making." *PLoS ONE* 9: e114013. https://doi.org/10.1371/journal.pone
.0114013.

6. A. H. Hastorf and H. Cantril. 1954. "They Saw a Game: A Case Study." *Journal of*

Abnormal and Social Psychology 49: 129–34. https://www.all-about-psychology
.com/selective-perception.html.

7. Ibid., 133.

8. Ibid.

9. Dan Kahan. May 4, 2011. "What Is Motivated Reasoning and How Does It Work?" *Science & Religion Today.* Accessed Nov. 22, 2019. http://www .scienceandreligiontoday.com/2011/05/04/what-is-motivated-reasoning-and -how-does-it-work/.

10. D. M. Kahan, E. Peters, E. C. Dawson, and P. Slovic. 2017. "Motivated Numeracy and Enlightened Self-Government." *Behavioural Public Policy* 1: 54–86. https://doi.org/10.1017/bpp.2016.2.

11. Ezra Klein. April 6, 2014. "How Politics Makes Us Stupid." Vox. https://www .vox.com/2014/4/6/5556462/brain-dead-how-politics-makes-us-stupid.

12. Rafi Letzter. June 4, 2018. "How Do DNA Ancestry Tests Really Work?" Live Science. https://www.livescience.com/62690-how-dna-ancestry-23andme -tests-work.html.

13. N. G. Crawford, D. E. Kelly, M. E. B. Hansen, M. H. Beltrame, S. Fan, S. L. Bowman, E. Jewett, et al. 2017. "Loci Associated with Skin Pigmentation Identified in African Populations." *Science* 358: eaan8433. https://doi.org/10.1126 /science.aan8433.

14. Belinda Luscombe. March 27, 2019. "What Police Departments and the Rest of Us Can Do to Overcome Implicit Bias, According to an Expert." *Time.* Accessed June 3, 2019. https://time.com/5558181/jennifer-eberhardt-overcoming -implicit-bias/. [[CQ: authors: supply month.]]

15. E. G. Bruneau, M. Cikara, and R. Saxe. 2015. "Minding the Gap: Narrative Descriptions about Mental States Attenuate Parochial Empathy." *PLoS ONE* 10: e0140838. https://doi.org/10.1371/journal.pone.0140838.

16. D. J. Kelly, P. C. Quinn, A. M. Slater, K. Lee, A. Gibson, M. Smith, L. Ge, et al. 2005. "Three-Month-Olds, but Not Newborns, Prefer Own-Race Faces." *Developmental Science* 8: F31–F36. https://doi.org/10.1111/j.1467-7687.2005 .0434a.x.

17. D. J. Kelly, P. C. Quinn, A. M. Slater, K. Lee, L. Ge, and O. Pascalis. 2007. "The Other-Race Effect Develops During Infancy: Evidence of Perceptual Narrowing." *Psychological Science* 18: 1084–89. https://doi.org/10.1111/j.1467 -9280.2007.02029.x.

18. Ibid.

19. S. Sangrigoli, C. Pallier, A.-M. Argenti, V. A. G. Ventureyra, and S. de Schonen. 2005. "Reversibility of the Other-Race Effect in Face Recognition During Childhood." *Psychological Science* 16: 440–44. https://doi.org/10.1111 /j.0956-7976.2005.01554.x.

20. D. S. Y. Tham, J. G. Bremner, and D. Hay. 2017. "The Other-Race Effect in Children from a Multiracial Population: A Cross-Cultural Comparison." *Journal of Experimental Child Psychology* 155: 128–37. https://doi.org/10.1016/j .jecp.2016.11.006.

21. C. Hughes, L. G. Babbitt, and A. C. Krendl. 2019. "Culture Impacts the Neural Response to Perceiving Outgroups Among Black and White Faces." *Frontiers in Human Neuroscience* 13. https://doi.org/10.3389/fnhum.2019.00143.

22. M. M. Davis, S. M. Hudson, D. S. Ma, and J. Correll. 2015. "Childhood Contact Predicts Hemispheric Asymmetry in Cross-Race Face Processing." *Psychonomic Bulletin & Review* 23: 824–30. https://doi.org/10.3758/s13423-015 -0972-7.

23. J. L. Eberhardt, P. A. Goff, V. J. Purdie, and P. G. Davies. 2004. "Seeing Black: Race, Crime, and Visual Processing." *Journal of Personality and Social Psychology* 87: 876–93. https://doi.org/10.1037/0022-3514.87.6.876.

24. Kerry Flynn. Feb. 8, 2019. "How Nextdoor Is Using Verified Location Data to Quietly Build a Big Ads Business." Digiday. https://digiday.com/marketing /nextdoor-ads-verified-homeowners/.
Lara O'Reilly. June 27, 2019. "How a Small Design Tweak Reduced Racial Profiling on Nextdoor by 75%" Yahoo! Finance. https://finance.yahoo.com/news /how-a-small-design-tweak-cut-racial-profiling-on-nextdoor-by-75-070000670 .html.

25. Nick Brinkerhoff, personal communication, Dec. 11, 2019.

26. J. W. Tanaka, B. Heptonstall, and S. Hagen. 2013. "Perceptual Expertise and the Plasticity of Other-Race Face Recognition." *Visual Cognition* 21: 1183– 1201. https://doi.org/10.1080/13506285.2013.826315.
J. W. Tanaka and L. J. Pierce. 2009. "The Neural Plasticity of Other-Race Face Recognition." *Cognitive, Affective, & Behavioral Neuroscience* 9: 122–31. https://doi.org/10.3758/CABN.9.1.122.

27. J. Y. Chiao, H. E. Heck, K. Nakayama, and N. Ambady. 2006. "Priming Race in Biracial Observers Affects Visual Search for Black And White

Faces." *Psychological Science* 17: 387–92. https://doi.org/10.1111/j.1467-9280 .2006.01717.x.

28. Matthew Frye Jacobson. 1998. *Whiteness of a Different Color: European Immigrants and the Alchemy of Race* (Cambridge, MA: Harvard University Press).

29. I. Minio-Paluello, S. Baron-Cohen, A. Avenanti, V. Walsh, and S. M. Aglioti. 2009. "Absence of Embodied Empathy During Pain Observation in Asperger Syndrome." *Biological Psychiatry* 65: 55–62. https://doi.org/10.1016/j .biopsych.2008.08.006.

30. Ibid.

31. J. Holroyd, R. Scaife, and T. Stafford. 2017. "Responsibility for Implicit Bias." *Philosophy Compass* 12: e12410. https://doi.org/10.1111/phc3.12410.

32. G. Hein, G. Silani, K. Preuschoff, C. D. Batson, and T. Singer. 2010. "Neural Responses to Ingroup and Outgroup Members' Suffering Predict Individual Differences in Costly Helping." *Neuron* 68: 149–60. https://doi.org/10.1016/j .neuron.2010.09.003.

33. Y. Dunham, A. S. Baron, and S. Carey. 2011. "Consequences of 'Minimal' Group Affiliations in Children." *Child Development* 82: 793–811. https://doi .org/10.1111/j.1467-8624.2011.01577.x.

34. B. Gaesser and D. L. Schacter. 2014. "Episodic Simulation and Episodic Memory Can Increase Intentions to Help Others." *Proceedings of the National Academy of Sciences* 111: 4415–20. https://doi.org/10.1073/pnas.1402461111.

35. C. D. Batson, J. Chang, R. Orr, and J. Rowland. 2002. "Empathy, Attitudes, and Action: Can Feeling for a Member of a Stigmatized Group Motivate One to Help the Group?" *Personality and Social Psychology Bulletin* 28: 1656–66. https://doi.org/10.1177/014616702237647.

36. Annette Gordon-Reed. June 6, 2011. "'Uncle Tom's Cabin' and the Art of Persuasion." *New Yorker.* https://www.newyorker.com/magazine/2011/06/13/the -persuader-annette-gordon-reed.

37. History.com Editors. Feb. 7, 2019. "Harriet Beecher Stowe." Accessed Dec. 14, 2019. https://www.history.com/topics/american-civil-war/harriet-beecher -stowe#section_3.

38. William Jennings Bryan, ed. "In the First Debate with Douglas by Abraham Lincoln," in *The World's Famous Orations.* Vol. IX: *America II* (New York: Funk and Wagnalls). Accessed Dec. 14, 2019. https://www.bartleby.com/268 /9/23.html.

39. D. Vollaro. 2009. "Lincoln, Stowe, and the 'Little Woman/Great War' Story: The Making, and Breaking, of a Great American Anecdote." *Journal of the Abraham Lincoln Association* 30(1): 18–24. https://quod.lib.umich.edu/j/jala/2629860.0030.104/—lincoln-stowe-and-the-little-womangreat-war-story-the-making?rgn=main;view=fulltext.

40. A. Kayaoğlu, S. Batur, and E. Aslıtürk. Nov. 2014. "The Unknown Muzafer Sherif." *The Psychologist* 27: 830–33. Accessed June 30, 2019. https://thepsychologist.bps.org.uk/volume-27/edition-11/unknown-muzafer-sherif.

41. David Shariatmadari. April 16, 2018. "A Real-life *Lord of the Flies:* The Troubling Legacy of the Robbers Cave Experiment." *Guardian*. https://www.theguardian.com/science/2018/apr/16/a-real-life-lord-of-the-flies-the-troubling-legacy-of-the-robbers-cave-experiment.

42. G. Perry. Nov. 2014. "The View from the Boys." *The Psychologist* 27: 834–37. https://thepsychologist.bps.org.uk/volume-27/edition-11/view-boys.

43. Ibid.

44. Sarah McCammon. Sept. 22, 2018. "The Cajun Navy: Heroes or Hindrances in Hurricanes?" NPR. https://www.npr.org/2018/09/22/650636356/the-cajun-navy-heroes-or-hindrances-in-hurricanes.
Miriam Markowitz. Dec. 7, 2017. "'We'll Deal with the Consequences Later': The Cajun Navy and the Vigilante Future of Disaster Relief." *GQ*. https://www.gq.com/story/cajun-navy-and-the-future-of-vigilante-disaster-relief.

45. David W. Moore. Sept. 24, 2001. "Bush Job Approval Highest in Gallup History." https://news.gallup.com/poll/4924/bush-job-approval-highest-gallup-history.aspx.

9. ACCULTURATING

1. P. Grosjean. 2014. "A History of Violence: The Culture of Honor and Homicide in the US South." *Journal of the European Economic Association* 12: 1285–1316. https://doi.org/10.1111/jeea.12096.

2. Richard E. Nisbett and Dov Cohen. 1996. *Culture of Honor: The Psychology of Violence in the South* (Boulder, CO: Westview Press).

3. R. E. Nisbett. 1993. "Violence and U.S. Regional Culture." *American Psychologist* 48: 441–49. https://doi.org/10.1037/0003-066X.48.4.441.

4. D. Cohen, R. E. Nisbett, B. F. Bowdle, and N. Schwarz. 1996. "Insult, Aggression, and the Southern Culture of Honor: An 'Experimental Ethnography.'"

Journal of Personality and Social Psychology 70: 945–60. https://doi.org/10 .1037/0022-3514.70.5.945.

5. S. Busatta. 2006. "Honour and Shame in the Mediterranean." *Antropologia Culturale* 2: 75–78. https://www.academia.edu/524890/Honour_and_Shame _in_the_Mediterranean.

6. Editors of *Encyclopedia Britannica*. "seppuku." *Encyclopedia Britannica*. Accessed July 1, 2019. https://www.britannica.com/topic/seppuku.

7. Elijah Anderson. 1999. *Code of the Street: Decency, Violence, and the Moral Life of the Inner City* (New York: W. W. Norton).

8. Drake Baer. July 14, 2016. "Gun Violence Is Like an STI in the Way It Moves Between People." https://www.thecut.com/2016/07/gun-violence-is-like-an-sti.html.

9. David Hackett Fischer. 1989. *Albion's Seed: Four British Folkways in America* (Oxford: Oxford University Press).

10. Richard E. Nisbett, personal interview, May 1, 2019.

11. Grosjean. 2014. "A History of Violence."

12. R. E. Nisbett, K. Peng, I. Choi, and A. Norenzayan. 2001. "Culture and Systems of Thought: Holistic Versus Analytic Cognition." *Psychological Review* 108: 291–310. https://doi.org/10.1037/0033-295X.108.2.291.

13. Ibid., 291.

14. Thomas Talhelm and Shigehiro Oishi. 2019. "Culture and Ecology," in *Handbook of Cultural Psychology,* ed. Dov Cohen and Shinobu Kitayama, 2nd ed. (New York: Guilford Press), 119–43.

15. L.-H. Chiu. 1972. "A Cross-Cultural Comparison of Cognitive styles in Chinese and American Children." *International Journal of Psychology* 7: 235–42. https://doi.org/10.1080/00207597208246604.

16. Paula Gottlieb. March 6, 2019. "Aristotle on Non-contradiction." *Stanford Encyclopedia of Philosophy,* ed. Edward N. Zalta. https://plato.stanford.edu /archives/spr2019/entries/aristotle-noncontradiction/.

17. Graham Priest, Francesco Berto, and Zach Weber. "Dialetheism." *Stanford Encyclopedia of Philosophy,* ed. Edward N. Zalta. Accessed July 1, 2019. https:// plato.stanford.edu/entries/dialetheism/

18. Richard E. Nisbett. 2003. *The Geography of Thought: How Asians and Westerners Think Differently . . . and Why* (New York: Free Press).

19. V. L. Vignoles, E. Owe, M. Becker, P. B. Smith, M. J. Easterbrook, R. Brown, R. González, et al. 2016. "Beyond The 'East–West' Dichotomy: Global Vari-

ation in Cultural Models of Selfhood." *Journal of Experimental Psychology: General* 145: 968. https://doi.org/10.1037/xge0000175.

20. Ibid.

21. T. Masuda and R. E. Nisbett. 2001. "Attending Holistically Versus Analytically: Comparing the Context Sensitivity of Japanese and Americans." *Journal of Personality and Social Psychology* 81: 922–34. https://doi.org/10.1037//0022 -3514.81.5.922.

22. Richard Nisbett, personal interview, May 3, 2019.

23. Nisbett et al. "Culture and Systems of Thought."

24. M. E. W. Varnum, I. Grossmann, S. Kitayama, and R. E. Nisbett. 2010. "The Origin of Cultural Differences in Cognition: The Social Orientation Hypothesis." *Current Directions in Psychological Science* 19: 9–13. https://doi.org/10 .1177/0963721409359301.

25. S. de Oliveira and R. E. Nisbett. 2017. "Beyond East and West: Cognitive Style in Latin America." *Journal of Cross-Cultural Psychology* 48: 1554–77. https:// doi.org/10.1177/0022022117730816.

26. U. Kühnen, B. Hannover, U. Roeder, A. A. Shah, B. Schubert, A. Upmeyer, and S. Zakaria. 2001. "Cross-Cultural Variations in Identifying Embedded Figures: Comparisons from the United States, Germany, Russia, and Malaysia." *Journal of Cross-Cultural Psychology* 32: 365–72. https://doi.org/10.1177 /0022022101032003007.

27. S. Kitayama, K. Ishii, T. Imada, K. Takemura, and J. Ramaswamy. 2006. "Voluntary Settlement and the Spirit of Independence: Evidence from Japan's 'Northern Frontier.'" *Journal of Personality and Social Psychology* 91: 369–84. https://psycnet.apa.org/doi/10.1037/0022-3514.91.3.369.

28. R. E. Nisbett and Y. Miyamoto. 2005. "The Influence of Culture: Holistic Versus Analytic Perception." *Trends in Cognitive Sciences* 9: 467–73. https://doi .org/10.1016/j.tics.2005.08.004.

29. A. K. Uskul, R. E. Nisbett, and S. Kitayama. 2008. "Ecoculture, Social Interdependence and Holistic Cognition: Evidence from Farming, Fishing and Herding Communities in Turkey." *Communicative & Integrative Biology* 1: 40– 41. https://doi.org/10.4161/cib.1.1.6649.

30. Drake Baer. Feb. 14, 2017. "Rich People Literally See the World Differently." The Cut. https://www.thecut.com/2017/02/how-rich-people-see-the-world -differently.html.

31. B. Morling, S. Kitayama, and Y. Miyamoto. 2002. "Cultural Practices Emphasize Influence in the United States and Adjustment in Japan." *Personality and Social Psychology Bulletin* 28: 311–23. http://dx.doi.org/10.1177/0146167202286003.

J. L. Tsai, F. F. Miao, E. Seppala, H. H. Fung, and D. Y. Yeung. 2007. "Influence and Adjustment Goals: Sources of Cultural Differences in Ideal Affect." *Journal of Personality and Social Psychology* 92: 1102–17. https://doi.org/10.1037/0022-3514.92.6.1102.

32. Thomas Talhelm, personal communication, May 10, 2019.

33. Michael Hurwitz. Feb. 27, 2014. "Stereotypes Chinese People Have About Themselves." *Yoyo Chinese* (blog). https://www.yoyochinese.com/blog/learn-chinese-china-regional-stereotypes.

34. Talhelm, personal communication, May 10, 2019.

35. T. Talhelm, X. Zhang, S. Oishi, C. Shimin, D. Duan, X. Lan, and S. Kitayama. 2014. "Large-scale Psychological Differences within China Explained by Rice Versus Wheat Agriculture." *Science* 344: 603–8. https://doi.org/10.1126/science.1246850.

36. A. Alesina, P. Giuliano, and N. Nunn. 2013. "On the Origins of Gender Roles: Women and the Plough." *Quarterly Journal of Economics* 128: 469–530. https://doi.org/10.1093/qje/qjt005.

37. T. Talhelm et al. "Large-Scale Psychological Differences within China Explained by Rice Versus Wheat Agriculture."

38. Colin Peebles Christensen. July 3, 2018. "China's Coffee War Is Heating Up." *China Economic Review.* https://chinaeconomicreview.com/chinas-coffee-war-is-heating-up/.

39. R. Thomson, M. Yuki, T. Talhelm, J. Schug, M. Kito, A. H. Ayanian, J. C. Becker, et al. 2018. "Relational Mobility Predicts Social Behaviors in 39 Countries and Is Tied to Historical Farming and Threat." *Proceedings of the National Academy of Sciences* 115: 7521–26. https://doi.org/10.1073/pnas.1713191115.

40. Facebook screen shot. https://osf.io/546dc/.

EPILOGUE: THE PATH IS MADE BY WALKING

1. Neil Shubin. 2008. *Your Inner Fish: A Journey into the 3.5-Billion-Year History of the Human Body* (New York: Pantheon Books).

2. S. S. Fisher, M. McGreevy, J. Humphries, and W. Robinett. 1987. "Virtual Environment Display System." *Proceedings of the 1986 Workshop on Interactive 3D Graphics,* 77–87. https://doi.org/10.1145/319120.319127.

3. M. Slater and M. V. Sanchez-Vives. Dec. 19, 2016. "Enhancing Our Lives with Immersive Virtual Reality." *Frontiers in Robotic and AI 3.* https://doi.org/10.3389/frobt.2016.00074.

4. R. Pausch, J. Snoody, R. Taylor, S. Watson, and E. Haseltine. Aug. 1996. "Disney's *Aladdin*: First Steps Towards Storytelling in Virtual Reality." *SIGGRAPH '96: Proceedings of the 23rd Annual Conference on Computer Graphics,* 193–203. https://doi.org/10.1145/237170.237257.

5. B. van der Hoort, A. Guterstam, and H. H. Ehrsson. 2011. "Being Barbie: The Size of One's Own Body Determines the Perceived Size of the World." *PLoS ONE* 6: e20195. https://doi.org/10.1371/journal.pone.0020195.

6. N. Yee and J. Bailenson. 2007. "The Proteus Effect: The Effect of Transformed Self-Representation on Behavior." *Human Communication Research* 33: 271–90. https://doi.org/10.1111/j.1468-2958.2007.00299.x.

7. N. Yee, J. N. Bailenson, and N. Ducheneaut. 2009. "The Proteus Effect: Implications of transformed Digital Self-Representation on Online and Offline Behavior." *Communication Research* 36: 285–312. https://doi.org/10.1177/0093650208330254.

8. M. Matamala-Gomez, T. Donegan, S. Bottiroli, G. Sandrini, M. V. Sanchez-Vives, and C. Tassorelli. 2019. "Immersive Virtual Reality and Virtual Embodiment for Pain Relief." *Frontiers in Human Neuroscience* 13: 279. https://doi.org/10.3389/fnhum.2019.00279.

9. T. C. Peck, S. Seinfeld, S. M. Aglioti, and M. Slater. 2013. "Putting Yourself in the Skin of a Black Avatar Reduces Implicit Racial Bias." *Consciousness and Cognition* 22: 779–87. http://doi.org/10.1016/j.concog.2013.04.016.

10. D. Banakou, P. D. Hanumanthu, and M. Slater. 2016. "Virtual Embodiment of White People in a Black Virtual Body Leads to a Sustained Reduction in Their Implicit Racial Bias." *Frontiers in Human Neuroscience* 10: 601. https://doi.org/10.3389/fnhum.2016.00601.

11. N. Salmanowitz. 2018. "The Impact of Virtual Reality on Implicit Racial Bias and Mock Legal Decisions." *Journal of Law and the Biosciences* 5: 174–203. https://doi.org/10.1093/jlb/lsy005.

12. D. Banakou, S. Kishore, and M. Slater. 2018. "Virtually Being Einstein Results in an Improvement in Cognitive Task Performance and a Decrease in Age Bias." *Frontiers in Psychology* 9: 917. https://doi.org/10.3389/fpsyg.2018.00917.

13. S. A. Osimo, R. Pizarro, B. Spanlang, and M. Slater. 2015. "Conversations Between Self and Self as Sigmund Freud—A Virtual Body Ownership Paradigm for Self Counselling." *Scientific Reports* 5: 13899. https://doi.org/10.1038/srep13899.

14. E. Twedt, R. M. Rainey, and D. R. Proffitt. 2019. "Beyond Nature: The Roles of Visual Appeal and Individual Differences in Perceived Restorative Potential." *Journal of Environmental Psychology* 65: 1–11. https://doi.org/10.1016/j.jenvp.2019.101322.

    R. S. Ulrich, C. Zimring, X. Zhu, J. Dubose, H-B. Seo, Y-S. Choi, X. Quan, et al. 2008. "A Review of the Research Literature on Evidence-Based Healthcare Design." *HERD: Health Environments Research & Design Journal* 1: 61–125. https://doi.org/10.1177/193758670800100306.

15. E. J. Langer and J. Rodin. 1976. "The Effects of Choice and Enhanced Personal Responsibility for the Aged: A Field Experiment in an Institutional Setting." *Journal of Personality and Social Psychology* 34: 191–98. https://doi.org/10.1037/0022-3514.34.2.191.

16. R. Held and A. Hein. 1963. "Movement-Produced Stimulation in the Development of Visually Guided Behavior." *Journal of Comparative and Physiological Psychology* 56: 872–76. https://doi.org/10.1037/h0040546.

17. Antonio Machado (1912). Campos de Castilla [Fields of Castile], Madrid: Renacimiento, translated by Stanley Appelbaum, Dover Publications, 2007.

BIBLIOGRAPHY

Abrams, D. M., and M. J. Panaggio. 2012. "A Model Balancing Cooperation and Competition Can Explain Our Right-Handed World and the Dominance of Left-Handed Athletes." *Journal of the Royal Society Interface* 9: 2718–22. https://doi .org/10.1098/rsif.2012.0211.

Adolph, K. E. 2000. "Specificity of Learning: Why Infants Fall over a Veritable Cliff." *Psychological Science* 11: 290–95. https://doi.org/10.1111/1467-9280.00258.

Ahn, W. Y., K. T. Kishida, X. Gu, T. Lohrenz, A. Harvey, J. R. Alford, K. B. Smith, et al. 2014. "Nonpolitical Images Evoke Neural Predictors of Political Ideology." *Current Biology* 24: 2693–99. https://doi.org/10.1016/j.cub.2014.09.050.

Alesina, A., P. Giuliano, and N. Nunn. 2013. "On the Origins of Gender Roles: Women and the Plough." *Quarterly Journal of Economics* 128: 469–530. https:// doi.org/10.1093/qje/qjt005.

Anderson, Elijah. 1999. *Code of the Street: Decency, Violence, and the Moral Life of the Inner City.* New York: W. W. Norton.

Axelrod, F. B., and G. Gold-von Simson. 2007. "Hereditary Sensory and Autonomic Neuropathies: Types II, III, and IV." *Orphanet Journal of Rare Diseases* 2: 1–12. https://doi.org/10.1186/1750-1172-2-39.

Bacon, F. T. 1979. "Credibility of Repeated Statements: Memory for Trivia." *Journal*

of Experimental Psychology: Human Learning and Memory 5: 241–52. https://doi
.org/10.1037/0278-7393.5.3.241.

Banakou, D., P. D. Hanumanthu, and M. Slater. 2016. "Virtual Embodiment of
White People in a Black Virtual Body Leads to a Sustained Reduction in Their
Implicit Racial Bias." *Frontiers in Human Neuroscience* 10: 601. https://doi.org
/10.3389/fnhum.2016.00601.

Banakou, D., S. Kishore, and M. Slater. 2018. "Virtually Being Einstein Results
in an Improvement in Cognitive Task Performance and a Decrease in Age Bias."
Frontiers in Psychology 9: 917. https://doi.org/10.3389/fpsyg.2018.00917.

Bastien, G. J., B. Schepens, P. A. Willems, and N. C. Heglund. 2005. "Energetics
of Load Carrying in Nepalese Porters." *Science* 308: 1755. https://doi.org/10.1126
/science.1111513.

Batson, C. D., J. Chang, R. Orr, and J. Rowland. 2002. "Empathy, Attitudes, and
Action: Can Feeling for a Member of a Stigmatized Group Motivate One to Help
the Group?" *Personality and Social Psychology Bulletin* 28: 1656–66. https://doi
.org/10.1177/014616702237647.

Baumeister, J.-C., G. Papa, and F. Foroni. 2016. "Deeper than Skin Deep—The Effect of Botulinum Toxin-A on Emotion Processing." *Toxicon* 118: 86–90. https://
doi.org/10.1016/j.toxicon.2016.04.044.

Bennett, D. A., J. A. Schneider, Y. Tang, S. E. Arnold, and R. S. Wilson. 2006.
"The Effect of Social Networks on the Relation between Alzheimer's Disease
Pathology and Level of Cognitive Function in Old People: A Longitudinal Cohort Study." *The Lancet Neurology* 5: 406–12. https://doi.org/10.1016/S1474
-4422(06)70417-3.

Berkman, L. F., and S. L. Syme. 1979. "Social Networks, Host Resistance, and Mortality: A Nine-Year Follow-up Study of Alameda County Residents." *American
Journal of Epidemiology* 109: 186–204. https://doi.org/10.1093/oxfordjournals
.aje.a112674.

Bhalla, M., and D. R. Proffitt. 1999. "Visual-Motor Recalibration in Geographical
Slant Perception." *Journal of Experimental Psychology: Human Perception and
Performance* 25: 1076–96. https://doi:10.1037/0096-1523.25.4.1076.

Bloom, P. 2006. "My Brain Made Me Do It." *Journal of Cognition and Culture* 6:
209–14. https://doi.org/10.1163/156853706776931303.

———. 2007. "Religion Is Natural." *Developmental Science* 10: 147–51. https://doi
.org/10.1111/j.1467-7687.2007.00577.x.

Brockmole, J. R., C. C. Davoli, R. A. Abrams, and J. K. Witt. 2013. "The World Within Reach: Effects of Hand Posture and Tool Use on Visual Cognition." *Current Directions in Psychological Science* 22: 38–44. https://doi.org/10.1177/0963721412465065.

Bruneau, E. G., M. Cikara, and R. Saxe. 2015. "Minding the Gap: Narrative Descriptions about Mental States Attenuate Parochial Empathy." *PLoS ONE* 10: e0140838. https://doi.org/10.1371/journal.pone.0140838.

Busatta, S. 2006. "Honour and Shame in the Mediterranean." *Antropologia Culturale* 2: 75–78. https://www.academia.edu/524890/Honour_and_Shame_in_the_Mediterranean.

Campos, J. J., B. I. Bertenthal, and R. Kermoian. 1992. "Early Experience and Emotional Development: The Emergence of Wariness of Heights." *Psychological Science* 3: 61–64. https://doi.org/10.1111/j.1467-9280.1992.tb00259.x.

Carota, F., R. Moseley, and F. Pulvermüller. 2012. "Body-Part-Specific Representations of Semantic Noun Categories." *Journal of Cognitive Neuroscience* 24: 1492–1509. https://doi.org/10.1162/jocn_a_00219.

Casasanto, D., and E. G. Chrysikou. 2011. "When Left Is 'Right': Motor Fluency Shapes Abstract Concepts." *Psychological Science* 22: 419–22. https://doi.org/10.1177/0956797611401755.

Casasanto, D., and A. de Bruin. 2019. "Metaphors We Learn By: Directed Motor Action Improves Word Learning." *Cognition* 182: 177–83. https://doi.org/10.1016/j.cognition.2018.09.015.

Casasanto, D., and K. Dijkstra. 2010. "Motor Action and Emotional Memory." *Cognition* 115: 179–85. https://doi.org/10.1016/j.cognition.2009.11.002.

Casasanto, D., and K. Jasmin. 2010. "Good and Bad in the Hands of Politicians: Spontaneous Gestures during Positive and Negative Speech." *PLoS ONE* 14: e11805. https://doi.org/10.1371/journal.pone.0011805.

Chiao, J. Y., H. E. Heck, K. Nakayama, and N. Ambady. 2006. "Priming Race in Biracial Observers Affects Visual Search for Black And White Faces." *Psychological Science* 17: 387–92. https://doi.org/10.1111/j.1467-9280.2006.01717.x.

Chiu, L.-H. 1972. "A Cross-Cultural Comparison of Cognitive styles in Chinese and American Children." *International Journal of Psychology* 7: 235–42. https://doi.org/10.1080/00207597208246604.

Clore, Gerald L. 2018. "The Impact of Affect Depends on its Object." In *The Nature of Emotion: Fundamental Questions*. Edited by Andrew S. Fox, Regina C.

Lapate, Alexander J. Shackman, and Richard J. Davidson. 2nd ed. Oxford: Oxford University Press, 188–89.

Clore, G. L., and J. Palmer. 2009. "Affective Guidance of Intelligent Agents: How Emotion Controls Cognition." *Cognitive Systems Research* 10: 21–30. https://doi.org/10.1016/j.cogsys.2008.03.002.

Coan, J. A., L. Beckes, M. Z. Gonzalez, E. L. Maresh, C. L. Brown, and K. Hasselmo. 2017. "Relationship Status and Perceived Support in the Social Regulation of Neural Responses to Threat." *Social Cognitive and Affective Neuroscience* 12: 1574–83. https://doi.org/10.1093/scan/nsx091.

Coan, J. A., and D. A. Sbarra. 2015. "*Social Baseline Theory*: The Social Regulation of Risk and Effort." *Current Opinion in Psychology* 1: 87–91. https://doi.org/10.1016/j.copsyc.2014.12.021.

Cohen, D., R. E. Nisbett, B. F. Bowdle, and N. Schwarz. 1996. "Insult, Aggression, and the Southern Culture of Honor: An 'Experimental Ethnography.'" *Journal of Personality and Social Psychology* 70: 945–60. https://doi.org/10.1037/0022-3514.70.5.945.

Cole, S., E. Balcetis, and D. Dunning. 2012. "Affective Signals of Threat Increase Perceived Proximity." *Psychological Science* 24: 34–40. https://doi.org/10.1177/0956797612446953.

Corballis, M. C. 1980. "Laterality and Myth." *American Psychologist* 284–95. http://dx.doi.org/10.1037/0003-066X.35.3.284.

Correll, J., B. Park, C. M. Judd, and B. Wittenbrink. 2007. "The Influence of Stereotypes on Decisions to Shoot." *European Journal of Social Psychology* 37: 1102–17. https://doi.org/10.1002/ejsp.450.

Cosman, J. D., and S. P. Vecera. 2010. "Attention Affects Visual Perceptual Processing Near the Hand." *Psychological Science* 58. https://doi.org/10.1177/0956797610380697.

Cottingham, John, ed. 1996. *René Descartes: Meditations on First Philosophy With Selections from the Objections and Replies.* 2nd ed. Cambridge: Cambridge University Press.

Crawford, J. T., Y. Inbar, and V. Maloney. 2014. "Disgust Sensitivity Selectively Predicts Attitudes Toward Groups That Threaten (or Uphold) Traditional Sexual Morality." *Personality and Individual Differences* 70: 218–23. https://doi.org/10.1016/j.paid.2014.07.001.

Crawford, N. G., D. E. Kelly, M. E. B. Hansen, M. H. Beltrame, S. Fan, S. L. Bowman, E. Jewett, et al. 2017. "Loci Associated with Skin Pigmentation Identified in African Populations." *Science* 358: eaan8433. https://doi.org/10.1126/science .aan8433.

Danziger, S., J. Levav, and L. Avnaim-Pesso. 2011. "Extraneous Factors in Judicial Decisions." *Proceedings of the National Academy of Sciences* 108: 6889–92. https:// doi.org/10.1073/pnas.1018033108.

Darwin, Charles. 1872. *The Origin of Species by Means of Natural Selection; or, The Preservation of Favored Races in the Struggle for Life and The Descent of Man and Selection in Relation to Sex*, 6th ed. Reprint, New York: Modern Library, 1936.

Davis, M. M., S. M. Hudson, D. S. Ma, and J. Correll. 2015. "Childhood Contact Predicts Hemispheric Asymmetry in Cross-Race Face Processing." *Psychonomic Bulletin & Review* 23: 824–30. https://doi.org/10.3758/s13423-015-0972-7.

Davoli, C. C., and J. Brockmole. 2012. "The Hands Shield Attention from Visual Interference." *Attention, Perception, & Psychophysics* 71: 1386–90. https://doi.org /10.3758/s13414-012-0351-7.

de Gelder, B., M. Tamietto, G. van Boxtel, R. Goebel, A. Sahraie, J. van den Stock, B. M. C. Stienen, et al. 2008. "Intact Navigation Skills after Bilateral Loss of Striate Cortex." *Current Biology* 18: R1128–29. https://doi.org/10.1016/j.cub.2008 .11.002.

Defazio, G., G. Abbruzzese, P. Girlanda, L. Vacca, A. Currà, R. De Salvia et al. 2002. "Botulinum Toxin A Treatment for Primary Hemifacial Spasm." *Archives of Neurology* 59: 418–20. https://doi.org/10.1001/archneur.59.3.418

de Oliveira, S., and R. E. Nisbett. 2017. "Beyond East and West: Cognitive Style in Latin America." *Journal of Cross-Cultural Psychology* 48: 1554–77. https://doi.org /10.1177/0022022117730816.

DeWall, C. N., R. F. Baumeister, M. T. Gailliot, and J. K. Maner. 2008. "Depletion Makes the Heart Grow Less Helpful: Helping as a Function of Self-Regulatory Energy and Genetic Relatedness." *Personality and Social Psychology Bulletin* 34: 1653–62. https://doi.org/10.1177/0146167208323981.

Doerrfeld, A., N. Sebanz, and M. Shiffrar. 2012. "Expecting to Lift a Box Together Makes the Load Look Lighter." *Psychological Research* 76: 467–75. https://doi.org /10.1007/s00426-011-0398-4.

Dreyer, F. R., and F. Pulvermüller. 2018. "Abstract Semantics in the Motor System?—An Event-Related fMRI Study on Passive Reading of Semantic Word Categories Carrying Abstract Emotional and Mental Meaning." *Cortex* 100: 52–70. https://doi.org/10.1016/j.cortex.2017.10.021.

Dunham, Y., A. S. Baron, and S. Carey. 2011. "Consequences of 'Minimal' Group Affiliations in Children." *Child Development* 82: 793–811. https://doi.org/10.1111/j.1467-8624.2011.01577.x.

Eberhardt, J. L., P. G. Davies, V. J. Purdie-Vaughns, and S. L. Johnson. 2006. "Looking Deathworthy: Perceived Stereotypicality of Black Defendants Predicts Capital-Sentencing Outcomes." *Psychological Science* 17: 383–86. https://doi.org/10.1111/j.1467-9280.2006.01716.x.

Eberhardt, J. L., P. A. Goff, V. J. Purdie, and P. G. Davies. 2004. "Seeing Black: Race, Crime, and Visual Processing." *Journal of Personality and Social Psychology* 87: 876–93. https://doi.org/10.1037/0022-3514.87.6.876.

Ein-Dor, T., and O. Tal. 2012. "Scared Saviors: Evidence that People High in Attachment Anxiety Are More Effective in Alerting Others to Threat." *European Journal of Social Psychology* 42: 667–71. https://doi.org/10.1002/ejsp.1895.

Eisenberger, N. I., M. D. Lieberman, and K. D. Williams. 2003. "Does Rejection Hurt? An fMRI Study of Social Exclusion." *Science* 302: 290–92. https://doi.org/10.1126/science.1089134.

Ertel, K. A., M. M. Glymour, and L. F. Berkman. 2009. "Social Networks and Health: A Life Course Perspective Integrating Observational and Experimental Evidence." *Journal of Social and Personal Relationships* 26: 73–92. https://doi.org/10.1177/0265407509105523.

Eves F. F. 2014. "Is There Any Proffitt in Stair Climbing? A Headcount of Studies Testing for Demographic Differences in Choice of Stairs." *Psychonomic Bulletin & Review* 21: 71–77. https://doi.org/10.3758/s13423-013-0463-7.

Faubert, J. 2013. "Professional Athletes Have Extraordinary Skills for Rapidly Learning Complex and Neutral Dynamic Visual Scenes." *Scientific Reports* 3: 1–3. https://doi.org/10.1038/srep01154.

Fazio, L. K., N. M. Brashier, B. K. Payne, and E. J. Marsh. 2015. "Knowledge Does Not Protect Against Illusory Truth." *Journal of Experimental Psychology: General* 144: 993–1002. https://doi.org/10.1037/xge0000098.

Ferry, A. L., S. J. Hespos, and S. R. Waxman. 2010. "Categorization in 3- and 4-Month-Old Infants: An Advantage of Words Over Tones." *Child Development* 81: 472–79. https://doi.org/10.1111/j.1467-8624.2009.01408.x.

Fischer, D., M. Messner, and O. Pollatos. 2017. "Improvement of Interoceptive Processes after an 8-Week Body Scan Intervention." *Frontiers in Human Neuroscience* 11: 452. https://doi.org/10.3389/fnhum.2017.00452.

Fischer, David Hackett. 1989. *Albion's Seed: Four British Folkways in America.* Oxford: Oxford University Press.

Fisher, S. S., M. McGreevy, J. Humphries, and W. Robinett. 1987. "Virtual Environment Display System." *Proceedings of the 1986 Workshop on Interactive 3D Graphics,* 77–87. https://doi.org/10.1145/319120.319127.

Flint, E., and S. Cummins. 2016. "Active Commuting and Obesity in Mid-Life: Cross-Sectional, Observational Evidence from UK Biobank." *The Lancet: Diabetes & Endocrinology* 4: 420–35. https://doi.org/10.1016/S2213-8587(16)00053-X.

Frankfurt, Harry G. 2005. *On Bullshit.* Princeton, NJ: Princeton University Press.

Gaesser, B., and D. L. Schacter. 2014. "Episodic Simulation and Episodic Memory Can Increase Intentions to Help Others." *Proceedings of the National Academy of Sciences* 111: 4415–20. https://doi.org/10.1073/pnas.1402461111.

Gallwey, W. Timothy. 1974. *The Inner Game of Tennis: The Classic Guide to the Mental Side of Peak Performance.* New York: Random House. Reprint, Toronto: Bantam Books, 1979.

Gibson, E. J., and R. D. Walk. 1960. "The 'Visual Cliff.'" *Scientific American* 202: 64–71. https://doi.org/10.1038/scientificamerican0460-64.

Gibson, James J. 1979. *The Ecological Approach to Visual Perception.* Boston: Houghton Mifflin.

Glenberg, A. M. 2011. "How Reading Comprehension Is Embodied and Why That Matters." *International Electronic Journal of Elementary Education* 4: 5–18. https://iejee.com/index.php/IEJEE/article/view/210.

Glenberg, A. M., M. Sato, L. Cattaneo, L. Riggio, D. Palumbo, and G. Buccino. 2008. "Processing Abstract Language Modulates Motor System Activity." *Quarterly Journal of Experimental Psychology* 61: 905–19. https://doi.org/10.1080/17470210701625550.

Glocker, M. L., D. D. Langleben, K. Ruparel, J. W. Loughead, R. C. Gur, and N. Sachser. 2009. "Baby Schema in Infant Faces Induces Cuteness Perception and Motivation for Caretaking in Adults." *Ethology* 115: 257–63. https://doi.org/10.1111/j.1439-0310.2008.01603.x.

Goldin-Meadow, S. 1997. "When Gestures and Words Speak Differently." *Current Directions in Psychological Science* 6: 138–43. https://doi.org/10.1111/1467-8721.ep10772905.

Goodale, Melvyn A., and David Milner. 2004. *Sight Unseen: An Exploration of Conscious and Unconscious Vision.* Oxford: Oxford University Press.

Gooriah, R., and F. Ahmed. 2015. "OnabotulinumtoxinA for Chronic Migraine: A Critical Appraisal." *Therapeutics and Clinical Risk Management* 11: 1003–13. https://doi.org/10.2147/TCRM.S76964.

Gray, R. 2013. "Being Selective at the Plate: Processing Dependence Between Perceptual Variables Relates to Hitting Goals and Performance." *Journal of Experimental Psychology: Human Perception and Performance* 39: 1124–42. https://doi.org/10.1037/a0030729.

Greenfield, Patricia Marks, and Joshua H. Smith. 1977. *The Structure of Communication in Early Language Development.* New York: Academic Press.

Grosjean, P. 2014. "A History of Violence: The Culture of Honor and Homicide in the US South." *Journal of the European Economic Association* 12: 1285–1316. https://doi.org/10.1111/jeea.12096.

Haffenden, A. M., K. C. Schiff, and M. A. Goodale. 2001. "The Dissociation Between Perception and Action in the Ebbinghaus Illusion: Nonillusory Effects of Pictorial Cues on Grasp." *Current Biology* 11: 177–81. https://doi.org/10.1016/S0960-9822(01)00023-9.

Harlow, H. F. 1958. "The Nature of Love." *American Psychologist* 13: 673–85.

Harnad, S. 1990. "The Symbol Grounding Problem." *Physica D: Nonlinear Phenomena* 42: 335–46. https://doi.org/10.1016/0167-2789(90)90087-6.

Hastorf, A. H., and H. Cantril. 1954. "They Saw a Game: A Case Study." *Journal of Abnormal and Social Psychology* 49: 129–34. https://www.all-about-psychology.com/selective-perception.html.

Hauk, O., I. Johnsrude, and F. Pulvermüller. 2004. "Somatotopic Representation of Action Words in Human Motor and Premotor Cortex." *Neuron* 41: 301–7. https://doi.org/10.1016/S0896-6273(03)00838-9.

Havas, D. A., A. M. Glenberg, K. A. Gutowski, M. J. Lucarelli, and R. J. Davidson. 2010. "Cosmetic Use of Botulinum Toxin-A Affects Processing of Emotional Language." *Psychological Science* 21: 895–900. https://doi.org/10.1177/0956797610374742.

Hein, G., G. Silani, K. Preuschoff, C. D. Batson, and T. Singer. 2010. "Neural Responses to Ingroup and Outgroup Members' Suffering Predict Individual Differences in Costly Helping." *Neuron* 68: 149–60. https://doi.org/10.1016/j.neuron.2010.09.003.

Held, R., and J. Bossom. 1961. "Neonatal Deprivation and Adult Rearrangement: Complementary Techniques for Analyzing Plastic Sensory-Motor Coordinations."

Journal of Comparative and Physiological Psychology 54: 33–37. https://doi.org/10 .1037/h0046207.

Held, R., and A. Hein. 1963. "Movement-Produced Stimulation in the Development of Visually Guided Behavior." *Journal of Comparative and Physiological Psychology* 56: 872–76. https://doi.org/10.1037/h0040546.

Holroyd, J., R. Scaife, and T. Stafford. 2017. "Responsibility for Implicit Bias." *Philosophy Compass* 12: e12410. https://doi.org/10.1111/phc3.12410.hein

Holt-Lunstad, J., T. B. Smith, and J. B. Layton. 2010. "Social Relationships and Mortality Risk: A Meta-analytic Review." *PLoS Medicine:* e1000316. https://doi .org/10.1371/journal.pmed.1000316.

Hrdy, Sarah Blaffer. 2005. "Comes the Child before Man: How Cooperative Breeding and Prolonged Postweaning Dependence Shaped Human Potential." In *Hunter-Gatherer Childhoods: Evolutionary, Developmental, and Cultural Perspectives.* Edited by Barry S. Hewlett and Michael E. Lamb. New York: Routledge.

———. 2009. *Mothers and Others: The Evolutionary Origins of Mutual Understanding.* Cambridge, MA: Harvard University Press.

———. 2016. "Development Plus Social Selection in the Emergence of 'Emotionally Modern' Humans." In *Childhood: Origins, Evolution, and Implication.* Edited by Courtney L. Meehan and Alyssa A. Crittenden. Albuquerque: University of New Mexico Press.

Hughes, C., L. G. Babbitt, and A. C. Krendl. 2019. "Culture Impacts the Neural Response to Perceiving Outgroups Among Black and White Faces." *Frontiers in Human Neuroscience* 13. https://doi.org/10.3389/fnhum.2019.00143.

Inbar, Y., D. A. Pizarro, and P. Bloom. 2009. "Conservatives Are More Easily Disgusted Than Liberals." *Cognition and Emotion* 23: 725. https://doi.org/10.1080 /02699930802110007.

———. 2012. "Disgusting Smells Cause Decreased Liking of Gay Men." *Emotion* 12: 23–27. https://doi.org/10.1037/a0023984.

Iverson, J. M., and S. Goldin-Meadow. 1998. "Why People Gesture When They Speak." *Nature* 396: 228. https://doi.org/10.1038/24300.

———. 2005. "Gesture Paves the Way for Language Development." *Psychological Science* 16: 367–71. https://doi.org/10.1111/j.0956-7976.2005.01542.x.

Jacobson, Matthew Frye. 1998. *Whiteness of a Different Color: European Immigrants and the Alchemy of Race.* Cambridge, MA: Harvard University Press.

Jefferies, E., K. Patterson, R. W. Jones, and M. A. Lambon Ralph. 2009. "Comprehension of Concrete and Abstract Words in Semantic Dementia." *Neuropsychology* 23: 492–99. https://doi.org/10.1037/a0015452.

Kahan, D. M., E. Peters, E. C. Dawson, and P. Slovic. 2017. "Motivated Numeracy and Enlightened Self-Government." *Behavioural Public Policy* 1: 54–86. https://doi.org/10.1017/bpp.2016.2.

Kahneman, Daniel, Paul Slovic, and Amos Tversky, eds. *Judgement under Uncertainty: Heuristics and Biases.* Cambridge: Cambridge University Press.

Kalesan, B., M. D. Villarreal, K. M. Keyes, and S. Galea. 2015. "Gun Ownership and Social Gun Culture." *Injury Prevention* 22: 216–20. https://doi.org/10.1136/injuryprev-2015-041586.

Kandasamy, N., S. N. Garfinkel, L. Page, B. Hardy, H. D. Critchley, M. Gurnell, and J. M. Coates. 2016. "Interoceptive Ability Predicts Survival on a London Trading Floor." *Scientific Reports* 6: 1–7. https://doi.org/10.1038/srep32986.

Karen, Robert. 1994. *Becoming Attached: First Relationships and How They Shape Our Capacity to Love.* New York: Warner Books. Reprint, Oxford: Oxford University Press, 1998.

Kelly, D. J., S. Liu, L. Ge, P. C. Quinn, A. M. Slater, K. Lee, Q. Liu, et al. 2007. "Cross-Race Preferences for Same-Race Faces Extend Beyond the African Versus Caucasian Contrast in 3-Month-Old Infants." *Infancy* 11: 87–95. https://doi.org/10.1080/15250000709336871.

Kelly, D. J., P. C. Quinn, A. M. Slater, K. Lee, L. Ge, and O. Pascalis. 2007. "The Other-Race Effect Develops During Infancy: Evidence of Perceptual Narrowing." *Psychological Science* 18 1084–89. https://doi.org/10.1111/j.1467-9280.2007.02029.x.

Kelly, D. J., P. C. Quinn, A. M. Slater, K. Lee, A. Gibson, M. Smith, L. Ge, et al. 2005. "Three-Month-Olds, but Not Newborns, Prefer Own-Race Faces." *Developmental Science* 8: F31–F36. https://doi.org/10.1111/j.1467-7687.2005.0434a.x

Kendon A. 2017. "Reflections on the 'Gesture-First' Hypothesis of Language Origins." *Psychonomic Bulletin & Review* 24: 163–70. https://doi.org/10.3758/s13423-016-1117-3.

Khalil, M., H. W. Zafar, V. Quarshie, and F. Ahmed. 2014. "Prospective Analysis of the Use of OnabotulinumtoxinA (BOTOX) in the Treatment of Chronic Migraine; Real-life Data in 254 Patients from Hull, UK." *Journal of Headache and Pain* 15: 1–9. https://doi.org/10.1186/1129-2377-15-54.

Kiken, L. G., N. J. Shook, J. L. Robins, and J. N. Clore. 2018. "Association between Mindfulness and Interoceptive Accuracy in Patients with Diabetes: Preliminary Evidence from Blood Glucose Estimates." *Complementary Therapies in Medicine* 36: 90–92. https://doi.org/10.1016/j.ctim.2017.12.003.

Kitayama, S., K. Ishii, T. Imada, K. Takemura, and J. Ramaswamy. 2006. "Voluntary Settlement and the Spirit of Independence: Evidence from Japan's 'Northern Frontier.'" *Journal of Personality and Social Psychology* 91: 369–84. https://psycnet.apa .org/doi/10.1037/0022-3514.91.3.369.

Klein, Colin. 2015. *What the Body Commands: The Imperative Theory of Pain.* Cambridge, MA: MIT Press.

Kringelbach, M. L., A. Lehtonen, S. Squire, A. G. Harvey, M. G. Craske, I. E. Holliday, A. L. Green, et al. 2008. "A Specific and Rapid Neural Signature for Parental Instinct." *PloS ONE* 3: e1664. https://doi.org/10.1371/journal.pone.0001664.

Króliczak, G., B. J. Piper, and S. H. Frey. 2011. "Atypical Lateralization of Language Predicts Cerebral Asymmetries in Parietal Gesture Representations." *Neuropsychologia* 49: 1698–1702. https://doi.org/10.1016/j.neuropsychologia .2011.02.044.

Kruger, A. C., and M. Konner. 2010. "Who Responds to Crying?: Maternal Care and Allocare among the !Kung." *Human Nature* 21: 309–29. https://doi.org/10 .1007/s12110-010-9095-z.

Kühnen, U., B. Hannover, U. Roeder, A. A. Shah, B. Schubert, A. Upmeyer, and S. Zakaria. 2001. "Cross-Cultural Variations in Identifying Embedded Figures: Comparisons from the United States, Germany, Russia, and Malaysia." *Journal of Cross-Cultural Psychology* 32: 365–72. https://doi.org/10.1177/0022022101032003007.

Land, Michael F., and Dan-Eric Nilsson. 2002. *Animal Eyes.* Oxford: Oxford University Press.

Langer, E. J., and J. Rodin. 1976. "The Effects of Choice and Enhanced Personal Responsibility for the Aged: A Field Experiment in an Institutional Setting." *Journal of Personality and Social Psychology* 34: 191–98. https://doi.org/10.1037 /0022-3514.34.2.191.

LeBarton, E. S., S. Goldin-Meadow, and S. Raudenbush. 2015. "Experimentally Induced Increases in Early Gesture Lead to Increases in Spoken Vocabulary." *Journal of Cognition and Development* 16: 199–220. https://doi.org/10.1080 /15248372.2013.858041.

Lee, Y., S. Lee, C. Carello, and M. T. Turvey. 2012. "An Archer's Perceived Form

Scales the 'Hitableness' of Archery Targets." *Journal of Experimental Psychology: Human Perception and Performance* 38: 125–31. https://doi.org/10.1037/a0029036.

Lehmann J., A. H. Korstjens, and R. I. M. Dunbar. 2007. "Group Size, Grooming and Social Cohesion in Primates." *Animal Behavior* 74: 1617–29. https://doi.org/10.1016/j.anbehav.2006.10.025.

Leibovich, T., N. Cohen, and A. Henik. 2016. "Itsy bitsy spider?: Valence and Self-Relevance Predict Size Estimation." *Biological Psychology* 121: 138–45. https://doi.org/10.1016/j.biopsycho.2016.01.009.

Levine, J. A., L. M. Lanningham-Foster, S. K. McCrady, A. C. Krizan, L. R. Olson, P. H. Kane, M. D. Jensen, et al. 2005. "Interindividual Variation in Posture Allocation: Possible Role in Human Obesity." *Science* 307: 584. https://doi.org/10.1126/science.1106561.

Libertus, K., A. S. Joh, and A. W. Needham. 2016. "Motor Training at 3 Months Affects Object Exploration 12 Months Later." *Developmental Science* 19: 1058–66. https://doi.org/10.1111/desc.12370.

Lieberman, D. E. 2011. "Four Legs Good, Two Legs Fortuitous: Brains, Brawn, and the Evolution of Human Bipedalism." In *In the Light of Evolution: Essays from the Laboratory and Field.* Edited by Jonathan B. Losos. Greenwood Village, CO: Roberts and Company.

Linkenauger, S. A., V. Ramenzoni, and D. R. Proffitt. 2010. "Illusory Shrinkage and Growth: Body-Based Rescaling Affects the Perception of Size." *Psychological Science* 21: 1318–25. https://doi.org/10.1177/0956797610380700.

Linkenauger, S. A., J. K. Witt, J. K. Stefanucci, J. Z. Bakdash, and D. R. Proffitt. 2009. "The Effects of Handedness and Reachability on Perceived Distance." *Journal of Experimental Psychology: Human Perception and Performance* 35: 1649–60. https://doi.org/10.1037/a0016875.

Lorenz, K. "*Der Kumpan in der Umwelt des Vogels. Der Artgenosse als auslösendes Moment sozialer Verhaltensweisen.*" *Journal für Ornithologie* 83: 137–215, 289–413.

MacDonald, G., and M. R. Leary. 2005. "Why Does Social Exclusion Hurt? The Relationship Between Social and Physical Pain." *Psychological Bulletin* 131: 202–23. https://doi.org/10.1037/0033-2909.131.2.202.

Masuda, T., and R. E. Nisbett. 2001. "Attending Holistically Versus Analytically:

Comparing the Context Sensitivity of Japanese and Americans." *Journal of Personality and Social Psychology* 81: 922–34. https://doi.org/10.1037//0022-3514.81 .5.922.

Matamala-Gomez, M., T. Donegan, S. Bottiroli, G. Sandrini, M. V. Sanchez-Vives, and C. Tassorelli. 2019. "Immersive Virtual Reality and Virtual Embodiment for Pain Relief." *Frontiers in Human Neuroscience* 13: 279. https://doi.org/10.3389 /fnhum.2019.00279.

McEnroe, John, with James Kaplan. 2002. *You Cannot Be Serious*. New York: G. P. Putnam's Sons. Reprint, London: Time Warner Paperbacks.

McGlone, M. S., and J. Tofighbakhsh. 1999. "The Keats Heuristic: Rhyme as Reason in Aphorism Interpretation." *Poetics* 26: 235–44. https://doi.org/10.1016 /S0304-422X(99)00003-0.

McNeill, David. 1996. *Hand and Mind: What Gestures Reveal about Thought*. New ed. Chicago: University of Chicago Press.

Minio-Paluello, I., S. Baron-Cohen, A. Avenanti, V. Walsh, and S. M. Aglioti. 2009. "Absence of Embodied Empathy During Pain Observation in Asperger Syndrome." *Biological Psychiatry* 65: 55–62. https://doi.org/10.1016/j.biopsych.2008 .08.006.

Moeller, B., H. Zoppke, and C. Frings. 2015. "What a Car Does to Your Perception: Distance Evaluations Differ from Within and Outside of a Car." *Psychonomic Bulletin & Review* 23: 781–88. https://doi.org/10.3758/s13423 -015-0954-9.

Morling, B., S. Kitayama, and Y. Miyamoto. 2002. "Cultural Practices Emphasize Influence in the United States and Adjustment in Japan." *Personality and Social Psychology Bulletin* 28: 311–23. http://dx.doi.org/10.1177 /0146167202286003.

Morrison, I., L. S. Löken, and H. Olausson. 2010. "The Skin as a Social Organ." *Experimental Brain Research* 204: 305–14. https://doi.org/10.1007/s00221-009 -2007-y.

Murrar, S., and M. Brauer. 2017. "Entertainment-Education Effectively Reduces Prejudice." *Group Processes & Intergroup Relations* 21: 1053–77. https://doi.org /10.1177/1368430216682350.

Needham, A., T. Barrett, and K. Peterman. 2002. "A Pick-me-up for Infants' Exploratory Skills: Early Simulated Experiences Reaching for Objects Using 'Sticky'

Mittens Enhances Young Infants' Object Exploration Skills." *Infant Behavior and Development* 25: 279–95. https://doi.org/10.1016/S0163-6383(02)00097-8.

Neisser, U. 1981. "Obituary: James J. Gibson (1904–1979)." *American Psychologist* 36: 214–5. https://doi.org/10.1037/h0078037.

Nietzsche, Friedrich. 2001.*The Gay Science: With a Prelude in German Rhymes and an Appendix of Songs.* Edited by Bernard Williams, translated by Josefine Nauckhoff, poems translated by Adrian Del Caro. Cambridge: Cambridge University Press.

Nisbett, R. E. 1993. "Violence and U.S. Regional Culture." *American Psychologist* 48: 441–49. https://doi.org/10.1037/0003-066X.48.4.441.

Nisbett, Richard E. 2004. *The Geography of Thought: How Asians and Westerners Think Differently . . . and Why.* New York: Free Press.

Nisbett, Richard E., and Dov Cohen. 1996. *Culture of Honor: The Psychology of Violence in the South.* Boulder, CO: Westview Press.

Nisbett, R. E., and Y. Miyamoto. 2005. "The Influence of Culture: Holistic Versus Analytic Perception." *Trends in Cognitive Sciences* 9: 467–73. https://doi.org/10.1016/j.tics.2005.08.004.

Nisbett, R. E., K. Peng, I. Choi, and A. Norenzayan. 2001. "Culture and Systems of Thought: Holistic Versus Analytic Cognition." *Psychological Review* 108: 291–310. https://doi.org/10.1037/0033-295X.108.2.291.

Nolen-Hoeksema, S. 2000. "The Role of Rumination in Depressive Disorders and Mixed Anxiety/Depressive Symptoms." *Journal of Abnormal Psychology* 109: 504–11. https://doi.org/10.1037/0021-843X.109.3.504.

Novack, M. A., E. L. Congdon, N. Hemani-Lopez, and S. Goldin-Meadow. 2014. "From Action to Abstraction: Using the Hands to Learn Math." *Psychological Science* 25: 903–10. https://doi.org/10.1177/0956797613518351.

Nowrasteh, A. "Terrorism and Immigration: A Risk Analysis." *Cato Institute Policy Analysis No. 798,* 1–26. https://papers.ssrn.com/sol3/papers.cfm?abstract_id=2842277.

Ohala, J. 1994. "The Frequency Code Underlies the Sound-Symbolic Use of Voice Pitch." In *Sound Symbolism.* Edited by Leanne Hinton, Johanna Nichols, and John J. Ohala. Cambridge: Cambridge University Press, 325–47. https://doi.org/10.1017/CBO9780511751806.022.

Olausson, H., Y. Lamarre, H. Backlund, C. Morin, B. G. Wallin, G. Starck, S. Ekholm, et al. 2002. "Unmyelinated Tactile Afferents Signal Touch and Project to Insular Cortex." *Nature Neuroscience* 5: 900–4. https://doi.org/10.1038/nn896.

Oppenheimer, D. M. 2008. "The Secret Life of Fluency." *Trends in Cognitive Sciences* 12: 237–41. https://doi.org/10.1016/j.tics.2008.02.014.

Oppezzo, M., and D. L. Schwartz. 2014. "Give Your Ideas Some Legs: The Positive Effect of Walking on Creative Thinking." *Journal of Experimental Psychology: Learning, Memory, and Cognition* 40: 1142–52. https://doi.org/10.1037/a0036577.

Osimo, S. A., R. Pizarro, B. Spanlang, and M. Slater. 2015. "Conversations Between Self and Self as Sigmund Freud—A Virtual Body Ownership Paradigm for Self Counselling." *Scientific Reports* 5: 13899. https://doi.org/10.1038/srep13899.

Panksepp, J., B. Herman, R. Conner, P. Bishop, and J. P. Scott. 1978. "The Biology of Social Attachments: Opiates Alleviate Separation Distress." *Biological Psychiatry* 13: 607–18.

Pausch, R., J. Snoody, R. Taylor, S. Watson, and E. Haseltine. Aug. 1996. "Disney's *Aladdin*: First Steps Towards Storytelling in Virtual Reality." *SIGGRAPH '96: Proceedings of the 23rd Annual Conference on Computer Graphics,* 193–203. https://doi.org/10.1145/237170.237257.

Peck, T. C., S. Seinfeld, S. M. Aglioti, and M. Slater. 2013. "Putting Yourself in the Skin of a Black Avatar Reduces Implicit Racial Bias." *Consciousness and Cognition* 22: 779–87. http://doi.org/10.1016/j.concog.2013.04.016.

Petrocelli, J. V. 2018. "Antecedents of Bullshitting." *Journal of Experimental Social Psychology* 76: 249–58. https://doi.org/10.1016/j.jesp.2018.03.004.

Phelps, E. A., S. Ling, and M. Carrasco. 2006. "Emotion Facilitates Perception and Potentiates the Perceptual Benefits of Attention." *Psychological Science* 17: 292–99. https://doi.org/10.1111/j.1467-9280.2006.01701.x.

Ping, R. M., S. Goldin-Meadow, and S. L. Beilock. 2014. "Understanding Gesture: Is the Listener's Motor System Involved?" *Journal of Experimental Psychology: General* 143: 196. https://doi.org/10.1037/a0032246.

Pinker, Steven. 2014. *The Sense of Style: The Thinking Person's Guide to Writing in the 21st Century.* London: Penguin Books.

Polansky, Ronald. 2007. *Aristotle's De Anima: A Critical Commentary.* Cambridge: Cambridge University Press.

Pontzer, Herman. Jan. 4, 2019. "Humans Evolved to Exercise." *Scientific American.* https://www.scientificamerican.com/article/humans-evolved-to-exercise/.

Proffitt, D. R., M. Bhalla, R. Gossweiler, and J. Midgett. 1995. "Perceiving Geographical Slant." *Psychonomic Bulletin & Review* 2: 409–48. https://doi.org/10.3758/BF03210980.

Rachman, S., and M. Cuk. 1992. "Fearful Distortions." *Behaviour Research and Therapy* 30: 583. https://doi.org/10.1016/0005-7967(92)90003-Y.

Rainville, P. 2002. "Brain Mechanisms of Pain Affect and Pain Modulation." *Current Opinion in Neurobiology* 12: 195–204. https://doi.org/10.1016/S0959-4388(02)00313-6.

Reed, C. L., R. Betz, J. P. Garza, and R. J. Roberts. 2010. "Grab It! Biased Attention in Functional Hand and Tool Space." *Attention, Perception, & Psychophysics* 72: 236–45. https://doi.org/10.3758/APP.72.1.236.

Riener, C. R., J. K. Stefanucci, D. R. Proffitt, and G. Clore. 2011. "An Effect of Mood on the Perception of Geographical Slant." *Cognition & Emotion* 25: 174–82. https://doi.org/10.1080/02699931003738026.

Rodkey, E. N. 2011. "The Woman Behind the Visual Cliff." *APA Monitor on Psychology* 42: 30. https://www.apa.org/monitor/2011/07-08/gibson.

Roper, S. O., and J. B. Yorgason. 2009. "Older Adults with Diabetes and Osteoarthritis and Their Spouses: Effects of Activity Limitations, Marital Happiness, and Social Contacts on Partners' Daily Mood." *Family Relations* 58: 460–74. https://doi.org/10.1111/j.1741-3729.2009.00566.x

Rusch, H. 2014. "The Evolutionary Interplay of Intergroup Conflict and Altruism in Humans: A Review of Parochial Altruism Theory and Prospects for Its Extension." *Proceedings of the Royal Society B: Biological Sciences* 281: 20141539. https://doi.org/10.1098/rspb.2014.1539.

Salmanowitz, N. 2018. "The Impact of Virtual Reality on Implicit Racial Bias and Mock Legal Decisions." *Journal of Law and the Biosciences* 5: 174–203. https://doi.org/10.1093/jlb/lsy005.

Sangrigoli, S., C. Pallier, A.-M. Argenti, V. A. G. Ventureyra, and S. de Schonen. 2005. "Reversibility of the Other-Race Effect in Face Recognition During Childhood." *Psychological Science* 16: 440–44. https://doi.org/10.1111/j.0956-7976.2005.01554.x.

Santana, E., and M. de Vega. 2011. "Metaphors Are Embodied, and So Are Their Literal Counterparts." *Frontiers in Psychology* 2. https://doi.org/10.3389/fpsyg.2011.00090.

Schmidtke, D. S., M. Conrad, and A. M. Jacobs. Feb. 12, 2014. "Phonological Iconicity." *Frontiers in Psychology* 5. https://doi.org/10.3389/fpsyg.2014.00080.

Schnall, S., J. Haidt, G. L. Clore, and A. H. Jordan. 2008. "Disgust as Embodied

Moral Judgment." *Personality and Social Psychology Bulletin* 34: 1096–1109. https://doi.org/10.1177/0146167208317771.

Schnall, S., K. D. Harber, J. K. Stefanucci, and D. R. Proffitt. 2008. "Social Support and the Perception of Geographical Slant." *Journal of Experimental Social Psychology* 44: 1246–55. https://doi.org/10.1016/j.jesp.2008.04.011.

Schwarz, N., H. Bless, F. Strack, G. Klumpp, H. Rittenauer-Schatka, and A. Simons. 1991. "Ease of Retrieval as Information: Another Look at the Availability Heuristic." *Journal of Personality and Social Psychology* 61: 195–202. https://doi.org/10.1037/0022-3514.61.2.195.

Schwarz, N., and G. L. Clore. 1983. "Mood, Misattribution, and Judgments of Well-Being: Informative and Directive Functions of Affective States." *Journal of Personality and Social Psychology* 45: 519–23. https://psycnet.apa.org/doi/10.1037/0022-3514.45.3.513.

Shubin, Neil. 2008. *Your Inner Fish: A Journey into the 3.5-Billion-year History of the Human Body.* New York: Pantheon Books.

Silk, J. B., S. C. Alberts, and J. Altmann. 2003. "Social Bonds of Female Baboons Enhance Infant Survival." *Science* 302: 1231–34. https://doi.org/10.1126/science.1088580.

Slater, M., and M. V. Sanchez-Vives. Dec. 19, 2016. "Enhancing Our Lives with Immersive Virtual Reality." *Frontiers in Robotic and AI* 3. https://doi.org/10.3389/frobt.2016.00074.

Sommers, S. R. 2006. "On Racial Diversity and Group Decision Making: Identifying Multiple Effects of Racial Composition on Jury Deliberations." *Journal of Personality and Social Psychology* 90: 597–612. https://doi.org/10.1037/0022-3514.90.4.597.

Spasojević, J., and L. B. Alloy. 2001. "Rumination as a Common Mechanism Relating Depressive Risk Factors to Depression." *Emotion* 1: 25–37. https://doi.org/10.1037//1528-3542.1.1.25.

Spitz, R. A. 1945. "Hospitalism: An Inquiry into the Genesis of Psychiatric Conditions in Early Childhood." *The Psychoanalytic Study of the Child* 1: 53–74. https://doi.org/10.1080/00797308.1945.11823126.

Steell, L., A. Garrido-Méndez, F. Petermann, X. Díaz-Martínez, M. A. Martínez, A. M. Leiva, C. Salas-Bravo, et al. 2018. "Active Commuting Is Associated with a Lower Risk of Obesity, Diabetes, and Metabolic Syndrome in Chilean

Adults." *Journal of Public Health* 40: 508–16, https://doi.org/10.1093/pubmed /fdx092.

Stefanucci, J. K., D. R. Proffitt, G. L. Clore, and N. Parekh. 2008. "Skating down a Steeper Slope: Fear Influences the Perception of Geographical Slant." *Perception* 37: 321–23. https://doi.org/10.1068/p5796.

Stefanucci, J. K., and J. Storbeck. 2009. "Don't Look Down: Emotional Arousal Elevates Height Perception." *Journal of Experimental Psychology: General* 138: 131–45. https://doi.org/10.1037/a0014797.

Sugovic, M., P. Turk, and J. K. Witt. 2016. "Perceived Distance and Obesity: It's What You Weigh, Not What You Think." *Acta Psychologica* 165: 1–8. https://doi .org/10.1016/j.actpsy.2016.01.012.

Talhelm, Thomas, and Shigehiro Oishi. 2019. "Culture and Ecology." In *Handbook of Cultural Psychology*. Edited by Dov Cohen and Shinobu Kitayama. 2nd ed. New York: Guilford Press, 119–43.

Talhelm, T., X. Zhang, S. Oishi, C. Shimin, D. Duan, X. Lan, and S. Kitayama. 2014. "Large-scale Psychological Differences within China Explained by Rice Versus Wheat Agriculture." *Science* 344: 603–8. https://doi.org/10.1126/science.1246850.

Tanaka, J. W., B. Heptonstall, and S. Hagen. 2013. "Perceptual Expertise and the Plasticity of Other-Race Face Recognition." *Visual Cognition* 21: 1183–1201. https://doi.org/10.1080/13506285.2013.826315.

Tanaka, J. W., and L. J. Pierce. 2009. "The Neural Plasticity of Other-Race Face Recognition." *Cognitive, Affective, & Behavioral Neuroscience* 9: 122–31. https:// doi.org/10.3758/CABN.9.1.122.

Taylor-Covill, G. A. H., and F. F. Eves. 2016. "Carrying a Biological 'Backpack': Quasi-Experimental Effects of Weight Status and Body Fat Change on Perceived Steepness." *Journal of Experimental Psychology: Human Perception and Performance* 42: 331–38. http://dx.doi.org/10.1037/xhp0000137.

Tham, D. S. Y., J. G. Bremner, and D. Hay. 2017. "The Other-Race Effect in Children from a Multiracial Population: A Cross-Cultural Comparison." *Journal of Experimental Child Psychology* 155: 128–37. https://doi.org/10.1016/j.jecp.2016 .11.006.

Thomson, R., M. Yuki, T. Talhelm, J. Schug, M. Kito, A. H. Ayanian, J. C. Becker, et al. 2018. "Relational Mobility Predicts Social Behaviors in 39 Countries and Is Tied to Historical Farming and Threat." *Proceedings of the National Academy of Sciences* 115: 7521–26. https://doi.org/10.1073/pnas.1713191115.

Tomasello, M., B. Hare, H. Lehmann, and J. Call. 2007. "Reliance on Head versus Eyes in the Gaze Following of Great Apes and Human Infants: The Cooperative Eye Hypothesis." *Journal of Human Evolution* 52: 314–20. https://doi.org/10.1016/j.jhevol.2006.10.001.

Tsai, J. L., F. F. Miao, E. Seppala, H. H. Fung, and D. Y. Yeung. 2007. "Influence and Adjustment Goals: Sources of Cultural Differences in Ideal Affect." *Journal of Personality and Social Psychology* 92: 1102–17. https://doi.org/10.1037/0022-3514.92.6.1102.

Twedt, E., R. M. Rainey, and D. R. Proffitt. 2019. "Beyond Nature: The Roles of Visual Appeal and Individual Differences in Perceived Restorative Potential." *Journal of Environmental Psychology* 65: 1–11. https://doi.org/10.1016/j.jenvp.2019.101322

Uchino, B. N. 2006. "Social Support and Health: A Review of Physiological Processes Potentially Underlying Links to Disease Outcomes." *Journal of Behavioral Medicine* 29: 377–87. https://doi.org/10.1007/s10865-006-9056-5.

Ulrich, R. S., C. Zimring, X. Zhu, J. Dubose, H-B. Seo, Y-S. Choi, X. Quan, et al. 2008. "A Review of the Research Literature on Evidence-Based Healthcare Design." *HERD: Health Environments Research & Design Journal* 1: 61–125. https://doi.org/10.1177/193758670800100306.

Uskul, A. K., R. E. Nisbett, and S. Kitayama. 2008. "Ecoculture, Social Interdependence and Holistic Cognition: Evidence from Farming, Fishing and Herding Communities in Turkey." *Communicative & Integrative Biology* 1: 40–41. https://doi.org/10.4161/cib.1.1.6649.

Vaccari, A. 2012. "Dissolving Nature: How Descartes Made us Posthuman." *Techné: Research in Philosophy and Technology* 16: 138–86. https://doi.org/10.5840/techne201216213.

Vainio, L., M. Schulman, K. Tiippana, and M. Vainio. 2013. "Effect of Syllable Articulation on Precision and Power Grip Performance." *PLoS ONE* 8: e53061. https://doi.org/10.1371/journal.pone.0053061.

van der Hoort, B., A. Guterstam, and H. H. Ehrsson. 2011. "Being Barbie: The Size of One's Own Body Determines the Perceived Size of the World." *PLoS ONE* 6: e20195. https://doi.org/10.1371/journal.pone.0020195.

Varnum, M. E. W., I. Grossmann, S. Kitayama, and R. E. Nisbett. 2010. "The Origin of Cultural Differences in Cognition: The Social Orientation Hypothesis." *Current Directions in Psychological Science* 19: 9–13. https://doi.org/10.1177/0963721409359301.

Vasey, M. W., M. R. Vilensky, J. H. Heath, C. N. Harbaugh, A. G. Buffington, and R. H. Fazio. 2012. "It was as big as my head, I swear!: Biased Spider Size Estimation in Spider Phobia." *Journal of Anxiety Disorders* 26: 20–24. https://doi.org/10.1016/j.janxdis.2011.08.009.

Vereijken, Beatrix, Karen Adolph, Mark Denny, Yaman Fadl, Simone Gill, and Ana Lucero. 1995. "Development of Infant Crawling: Balance Constraints on Interlimb Coordination." In *Studies in Perception and Action III*. Edited by Benoît G. Bardy, Reinould J. Bootsma, and Yves Guiard. New Jersey: Lawrence Erlbaum Associates, 255–58.

Vignoles, V. L., E. Owe, M. Becker, P. B. Smith, M. J. Easterbrook, R. Brown, R. González, et al. 2016. "Beyond The 'East–West' Dichotomy: Global Variation in Cultural Models of Selfhood." *Journal of Experimental Psychology: General* 145: 968. https://doi.org/10.1037/xge0000175.

Vollaro, D. 2009. "Lincoln, Stowe, and the 'Little Woman/Great War' Story: The Making, and Breaking, of a Great American Anecdote." *Journal of the Abraham Lincoln Association* 30(1): 18–24. https://quod.lib.umich.edu/j/jala/2629860.0030.104/—lincoln-stowe-and-the-little-womangreat-war-story-the-making?rgn=main;view=fulltext.

Walker, I., and C. Hulme. 1999. "Concrete Words Are Easier to Recall Than Abstract Words: Evidence for a Semantic Contribution to Short-Term Serial Recall." *Journal of Experimental Psychology: Learning, Memory, and Cognition* 25: 1256–71. https://doi.org/10.1037/0278-7393.25.5.1256.

Wang, X. T., and R. D. Dvorak. 2010. "Sweet Future: Fluctuating Blood Glucose Levels Affect Future Discounting." *Psychological Science* 21: 183–88. https://doi.org/10.1177/0956797609358096.

Watkins, E. R. 2008. "Constructive and Unconstructive Repetitive Thought." *Psychological Bulletin* 134: 163–206. https://doi.org/10.1037/0033-2909.134.2.163.

Watkins, E., N. J. Moberly, and M. L. Moulds. 2008. "Processing Mode Causally Influences Emotional Reactivity: Distinct Effects of Abstract Versus Concrete Construal on Emotional Response." *Emotion* 8: 364–78. https://doi.org/10.1037/1528-3542.8.3.364.

Wedekind, C., T. Seebeck, F. Bettens, and A. J. Paepke. 1995. "MHC-Dependent Mate Preferences in Humans." *Proceedings of the Royal Society B: Biological Sciences* 260: 245–49. https://doi.org/10.1098/rspb.1995.0087.

Weiskrantz, L., E. K. Warrington, M. D. Sanders, and J. Marshall. 1974. "Visual Capacity in the Hemianopic Field Following a Restricted Occipital Ablation." *Brain* 97: 709–28. https://doi.org/10.1093/brain/97.1.709.

Wesp, R., P. Cichello, E. B. Gracia, and K. Davis. 2004. "Observing and Engaging in Purposeful Actions with Objects Influences Estimates of Their Size." *Perception & Psychophysics* 66: 1261–67. https://doi.org/10.3758/bf03194996.

Willems, R. M., I. Toni, P. Hagoort, and D. Casasanto. 2009. "Body-Specific Motor Imagery of Hand Actions: Neural Evidence from Right- and Left-Handers." *Frontiers in Human Neuroscience* 3. https://doi.org/10.3389/neuro.09.039.2009.

Witt, J. K., and J. R. Brockmole. 2012. "Action Alters Object identification: Wielding a Gun Increases the Bias to See Guns." *Journal of Experimental Psychology: Human Perception and Performance* 38: 1159–67. https://doi.org/10.1037/a0027881.

Witt, J. K., and T. E. Dorsch. 2009. "Kicking to Bigger Uprights: Field Goal Kicking Performance Influences Perceived Size." *Perception* 38: 1328–40. https://doi.org/10.1068/p6325.

Witt, J. K., S. A. Linkenauger, J. Z. Bakdash, and D. R. Proffitt. 2008. "Putting to a Bigger Hole: Golf Performance Relates to Perceived Size." *Psychonomic Bulletin & Review* 15: 581–85. https://doi.org/10.3758/pbr.15.3.581.

Witt, J. K., and D. R. Proffitt. 2005. "See the Ball, Hit the Ball: Apparent Ball Size Is Correlated with Batting Average." *Psychological Science* 16: 937–39. https://doi.org/10.1111/j.1467-9280.2005.01640.x.

Witt, J. K., D. R. Proffitt, and W. Epstein. 2005. "Tool Use Affects Perceived Distance, But Only When You Intend to Use It." *Journal of Experimental Psychology: Human Perception and Performance* 31: 880–88. https://doi.org/10.1037/0096-1523.31.5.880.

Witt, J. K., D. M. Schuck, and J. E. T. Taylor. 2011. "Action-Specific Effects Underwater." *Perception* 40: 530–37. https://doi.org/10.1068/p6910.

Witt, J. K., and M. Sugovic. 2010. "Performance and Ease Influence Perceived Speed." *Perception* 39: 1341–53. https://doi.org/10.1068/p6699.

———. 2013. "Spiders Appear to Move Faster Than Non-Threatening Objects Regardless of One's Ability to Block Them." *Acta Psychologica* 143: 284–91. https://doi.org/10.1016/j.actpsy.2013.04.011.

Wittgenstein, Ludwig. 1953. *Philosophical Investigations*. Translated by G. E. M. Anscombe. Oxford, UK: Blackwell.

Wong, Kate, Nov. 24, 2014. "40 Years After Lucy: The Fossil That Revolutionized the Search for Human Origins." *Observations* (blog), *Scientific American*.

https://blogs.scientificamerican.com/observations/40-years-after-lucy-the-fossil
-that-revolutionized-the-search-for-human-origins/.

Yee, N., and J. Bailenson. 2007. "The Proteus Effect: The Effect of Transformed Self-Representation on Behavior." *Human Communication Research* 33: 271–90. https://doi.org/10.1111/j.1468-2958.2007.00299.x.

Yee, N., J. N. Bailenson, and N. Ducheneaut. 2009. "The Proteus Effect: Implications of transformed Digital Self-Representation on Online and Offline Behavior." *Communication Research* 36: 285–312. https://doi.org/10.1177/0093650208330254.

Yorgason, J. B., S. O. Roper, J. G. Sandberg, and C. A. Berg. 2012. "Stress Spill-over of Health Symptoms from Healthy Spouses to Patient Spouses in Older Married Couples Managing Both Diabetes and Osteoarthritis." *Families, Systems, & Health* 30: 330–43. https://doi.org/10.1037/a0030670.

Zadra, J. R., A. L. Weltman, and D. R. Proffitt. 2016. "Walkable Distances Are Bio-energetically Scaled." *Journal of Experimental Psychology: Human Perception and Performance* 42: 39–51. https://doi.org/10.1037/xhp0000107.

INDEX

INDEX

INDEX

INDEX